THE AUDITORY SYSTEM IN SLEEP

THE AUDITORY SYSTEM IN SLEEP

Second Edition

RICARDO A. VELLUTI
Centro de Medicina del Sueño, Facultad de Medicina,
Universidad CLAEH, Punta del Este, Uruguay

Academic Press is an imprint of Elsevier
125 London Wall, London EC2Y 5AS, United Kingdom
525 B Street, Suite 1800, San Diego, CA 92101-4495, United States
50 Hampshire Street, 5th Floor, Cambridge, MA 02139, United States
The Boulevard, Langford Lane, Kidlington, Oxford OX5 1GB, United Kingdom

British Library Cataloguing-in-Publication Data
A catalogue record for this book is available from the British Library

Library of Congress Cataloging-in-Publication Data
A catalog record for this book is available from the Library of Congress

ISBN: 978-0-12-810476-7

For Information on all Academic Press publications
visit our website at https://www.elsevier.com/books-and-journals

Publisher: Nikki Levy
Acquisition Editor: Melanie Tucker
Editorial Project Manager: Kathy Padilla
Production Project Manager: Poulouse Joseph
Cover Designer: Alan Studholme

Typeset by MPS Limited, Chennai, India

This book is dedicated to my closest family: my wife Marisa, my daughters Rosina and Francesca, and my sons Alejandro and Federico, as well as my three grandchildren, Valentín, Julieta and Indali.

CONTENTS

FOREWORD

Professor Ricardo Velluti is a scientist whose scholarship is based on vast scientific culture and successful experimental research. Early in his scientific career he took an interest in sleep and auditory physiology. This combination of topics brought about a deep insight into the mechanisms of information processing during the different behavioural states of the wake–sleep ultradian cycle. There is a particular aspect of auditory processing that deserves attention with respect to sleep. Namely, the fact that, among the many sensory channels informing the brain, the auditory one is open not only during wakefulness but also during sleep. The importance of this auditory peculiarity may be appreciated by considering that the auditory input is the most important information underlying the telereceptive interaction with the environment during sleep. From a general evolutionary viewpoint, it is obvious that this remarkable property of the auditory system entails survival relevance in many animal species.

On this basis, the research carried out in Velluti's laboratory concerned the influence of acoustic stimulation on sleep behaviour and the mechanisms of neuronal processing of acoustic information along the complex auditory pathways of the central nervous system during the ultradian wake–sleep cycle. Leaving the appreciation of this book's content to the reader, I wish to add a personal note regarding Ricardo Velluti's work. Our long-lasting collaboration allowed me to witness his full dedication to science and continuous striving for further progress in his field of study. His experimental cleverness and cultural scholarship put him high in writing a novel approach to the modulation and processing of auditory inputs in different behavioural states. In conclusion, I believe that his attractive and up-to-date book will be very useful to a large audience of readers including not only basic and clinical scientists but also students eager to expand their knowledge of new scientific developments in sleep and auditory physiology.

Professor Pier Luigi Parmeggiani
Bologna, Italy
14 June 2007

ACKNOWLEDGEMENTS

It is clear to me that I am indebted to many people along my way through the study of neuroscience. First of all is my mentor, with whom I really began to think, Prof. Elio García-Austt. We were engaged in many projects in Montevideo and afterward in Madrid. We were very good friends, sharing many episodes of our lives, the good and the bad ones. I will always remember our sailboat trips through the dangerous Río de la Plata waters.

Prof. Raúl Hernández-Peón, the brilliant Mexican neurophysiologist, was also part of my endeavor, when working with him at the Tlalpan laboratory in México DF, where I wrote my first paper on sleep (1963). We met later at New Haven and that was the last time I saw him, because he died some time later in an automobile crash. He opened the study of sleep for me, a subject that has followed me since the Prof. R. Hess death. At that time I realized that sleep was part of neuroscience. He inspired me to continue, up to now.

During 1966–67 I was an NIH postdoctoral fellow with Prof. Robert Galambos at Yale University. This a very significant stay because of Galambos' scientific approach to the neuroscience of that time and because we also had personal exchanges on many subjects. Many decades later we met again and continued to talk about the auditory efferent system, as if resuming a talk that we started sometime before.

During a Latin American Physiology Congress in México City, Prof. Carmine Clemente invited me to collaborate with him at the Anatomy Department—Brain Research at UCLA, where I arrived with a IBRO/UNESCO fellowship. It was a real great time to be at Los Angeles (1970–71). Sleep was then a prima donna of neuroscience; I have profited a lot from knowing and talking with him and other salient researchers working at UCLA. We met some time ago at his office, still in UCLA, remembering those times spent in close collaboration.

Second, I am indebted to my collaborators at the Neurophysiology Laboratory of the Facultad de Medicina, Universidad de la República. All of my colleagues there are cited in the References of each chapter of this book, because everyone coming to my laboratory was a *potential* researcher.

The two most important were José L. Peña and Marisa Pedemonte. Both contributed a lot to the laboratory and my personal development because of the many discussions and comments that came out every day. Prof. José Luis Peña is now at the Dominick P. Purpura Department of Neuroscience, Albert Einstein College of Medicine, New York, and Marisa Pedemonte is now Professor of Physiology at CLAEH University in Punta del Este, Uruguay.

Marisa is also my great companion on the adventure that is neuroscience, my wife, and the mother of our two girls, Francesca and Rossina, finally my everything up to the present time. I am also grateful to her for her collaboration during the editing of this book.

Two great scientist collaborators and friends of more than 20 years come now to my mind, Pier Luigi Parmeggiani and Peter M. Narins. Both were quite important in my scientific life and in common life as well. Both gave me their open friendship, which I continuously enjoy, because we still exchange scientific and, most important, nonscientific, ideas, visits, congress encounters, and many other activities, in Montevideo, Los Angeles, and Bologna, Italy.

GENERAL INTRODUCTION

The brain is a very complex information processing device, whatever the information it may be working on. The sensory input represents the whole fan of information the central nervous system (CNS) receives to elicit, after complex processing, its output responses, such as motor, endocrine, neurovegetative, behavioural activities or changes in the CNS capacities such as memory, learning, and so on. The information coming from the outer and the inner worlds during life is a meaningful influence on the brain phenotypic development and, in my particular topic, on sleep organization. An important purpose of the brain evolution is to allow the organism to properly interact with the environments, the external and the internal one (the body). In early developmental stages, from phylogenetic and ontogenetic viewpoints, the sensory information constitutes a relevant drive that controls the brain function and the general physiology in many ways. The development of each brain is genetically conditioned although a germane component is the continuous information coming in through the senses from both worlds, a phenomenon that continues throughout life; it is an endless process.

Since the sensory information in general is continuously reaching the CNS, its processing will be differentiated according to the current physiological state of the brain, during wakefulness; sleep stages I–IV; and, paradoxical sleep. An important point that should be added is that the brain itself can condition its own sensory input by controlling all receptors and nuclei through the sensory efferent systems, which are present in every incoming pathway. Therefore, by using this feedback possibility, the complex processing circuit may be completed through a functional 'closed-loop' system.

The natural light–dark sequence, phylogenetically archaic information, through the light receptor and its processing system, profoundly influences the sleep–wakefulness cycle. The circadian rhythm of melatonin, produced in most organisms from algae to mammals, is generated in the latter by a central pacemaker located in the suprachiasmatic nuclei of the hypothalamus, largely synchronized by cues from the light–dark cycle. Since the beginning of life, the brain and sensory systems complexity are in constant and mutual enrichment from both anatomical and functional perspectives. The auditory, olfactory, vestibular, and somaesthetic systems also developed, introducing more sensory data that progressively shaped a brain that began to reach its completion, leading to a dynamic end: the genetically established sleep–waking cycle features.

Early in the 20th century, the concept of sleep as the result of a blockage of the auditory inflow was introduced; later on, it was proposed that the extensive deafferentation of ascending sensory impulses to the isolated brain resulted in sleep. In any case, I am now putting forward that sensory information about the environment and the body continuously modulates the CNS activity during the sleep–wakefulness cycle.

Many inputs participated in modulating the waking life and influencing sleep, which led to postulate that wakefulness is supported by the sensory systems and that a lack or decreased level in their activity would lead to sleep, constituting the 'passive' sleep theory. In addition, interactions between sleep and sensory input in general were reported, and a surgical quasitotal deafferentation revealed significant changes in the characteristics of sleep and wakefulness.

Although profoundly modified, the processing of sensory information is still present during sleep. While all sensory systems show some influence on sleep, they are also reciprocally modulated by the sleep or waking state of the brain. Thus, the incoming sensory information may alter sleep and waking physiology; and conversely, the sleeping brain imposes rules on information processing. Although we do not yet completely understand how the brain processes sensory information, it is currently accepted that neuronal networks/cell assemblies can change depending on the information they receive throughout life.

Early in human life, auditory data could be recorded. A never published experimental approach carried out during 1966 at the Perinatology Centre of the University Hospital in Montevideo, showed the presence of auditory evoked potentials at very early stages in human life (Velluti, unpublished result). During a human mature foetus delivery, when the uterus dilation was about 5 cm, two regular EEG sterilized electrodes were attached to the foetus scalp by the obstetrician in charge. Through a small and shielded loudspeaker, clicks were delivered over the mother's womb surface while recording a clearcut averaged foetus auditory evoked potential using a CAT computer.

Why the auditory system? Several reasons support the notion of relating sleep to auditory physiology.

First, hearing is the only telereceptive sensory modality relatively open during sleep in microosmatic animals, acting as a continuous monitor of the environment, such as predator detection or a baby's cry during the night that awakens the parents.

Second, the auditory system has a conspicuous efferent component that makes it unique, featuring a complex anatomy located in parallel to the classic ascending pathway and functioning as an input controller particularly through the action of its most peripheral sites, the olivocochlear system.

Third, the auditory stimuli can affect human sleep; for example, a noisy night will reduce the total sleep time and be followed by sleep normalization once the noise is reduced.

Fourth, the total lack of auditory input after bilateral surgical lesion of the cochleae alters the sleep architecture of guinea pigs and hamsters by increasing the total sleep time.

Fifth, the link between auditory memory traces and sleep is also expressed by the presence of auditory images in 65% of recalled dreams.

Sixth, the local blood flow is significantly increased in auditory loci, such as the auditory cortex, medial geniculate nucleus, inferior colliculus, superior olive, and particularly, the cochlear nucleus.

Seventh, imaging using functional magnetic resonance showed that the presentation of auditory stimuli in slow wave sleep elicited a significant bilateral activation in the auditory, parietal, and prefrontal cortices as well as thalamus and cingulate.

Eighth, human and animal auditory evoked potentials as well as evoked magnetoencephalographic activity change waveforms and amplitudes on passing to sleep. Another electrophysiological approach, such as auditory single unit recording during sleep, also exhibited major shifts on passing to sleep (Chapter 5: Auditory Unit Activity in Sleep).

Including the auditory information processing during sleep is a step forward towards one of the probably main function taking place in sleep, memory storage and probably some kind of sleep learning. These high functions may be performed during sleep are part of the scenery of this still enigmatic behavioural condition.

This book begins with two chapters aiming at providing information about the two main approaches and trying to expose basic ideas on both the auditory system and the sleep behavioural state. It is not usual to analyse how the sensory information works during sleep because of many nonscientific prejudices. The brain is just one and everything that takes places within it must have influences on several other systems, such as the sleep and auditory functional relationships.

The third and fourth chapters are engaged in notes about information processing and neuronal networks intending to introduce us in a general and mainly theoretical approach to our problem.

The next chapter, the fifth, is a general view of the many experimental ways and results to analyse the sleep-auditory system relationship. Evoked potentials, local-field and far-field human recorded potentials, magnetoencephalography, as well as imaging are different points of interest exposed.

Chapter 5, Auditory Unit Activity in Sleep, approaches the problem of auditory processing by showing single units, their firings, their firing patterns, and their relationship with a well-known brain rhythm, the hippocampal theta rhythm.

The possible auditory influences on sleep are part of Chapter 6, Auditory Influences on Sleep. The auditory system constitutes an important way of incoming information that is not close during sleep, and that is why it can be altered by sound.

Finally, some conclusions will be made from an auditory viewpoint, from a sleep point of view and from a complex sleep-auditory standpoint.

Ricardo A. Velluti
ricardo.velluti@gmail.com

CHAPTER 1

Brief Analysis of the Auditory System Organization and its Physiologic Basis

The auditory system with its associated anatomical and functional complexity serves diverse processes, such as discrimination of sound frequencies and intensities, sound source location in space, auditory learning, development of human language and auditory 'images' in dreams, music and development of birds songs, that is, communication in general. In this chapter, a review of the known auditory ascending and descending systems are presented along with some new or not too well-known approaches.

THE AFFERENT ASCENDING SYSTEM

This complex system begins at the receptors in the cochlea followed by a wide upward expansion throughout the different nuclei, reticular formation, cerebellum and connections to the primary and secondary cortices. It is composed of several neuronal groups with profuse communication from the cochlea to the cortex.

Moreover, a nonclassical pathway assumed to branch off from classical path at the midbrain through connections from the central nucleus of the inferior colliculus (IC) (Moller and Rollins, 2002).

A diagram of the most important pathways and synaptic stations of the afferent auditory system is shown in Fig. 1.1 and a schematic one in Fig. 1.2. The first-order auditory neurons, with cell bodies located in Corti's ganglion, send their axons centrally to form the auditory nerve, part of the VIIIth cranial pair. These nerve fibres synapse with the secondary neurons located centrally in different cochlear nucleus (CN) *loci*, in the medulla–pontine region. Let us bear in mind that 95% of the

The Auditory System in Sleep
DOI: https://doi.org/10.1016/B978-0-12-810476-7.00001-4

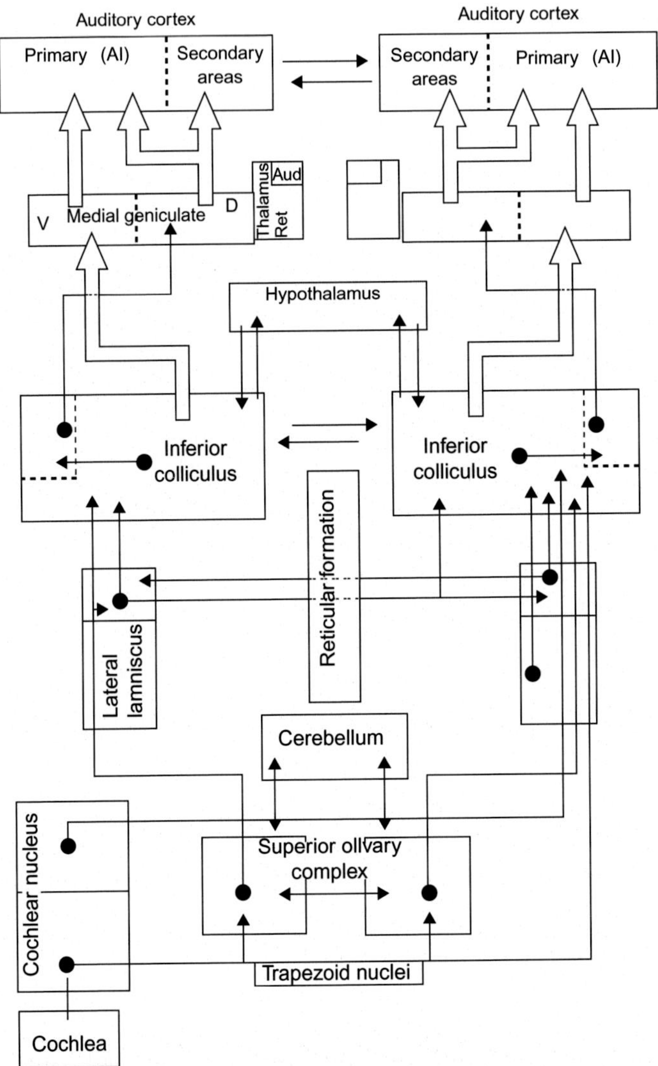

Figure 1.1 General diagram of the ascending auditory pathways. These pathways are embedded in the brain, an important concept since it reflects the multiple communication pathways that may affect incoming auditory information.

fibres that form the auditory nerve originate at the inner hair cells. The outer hair cells are innervated by only 5%, nonmyelinated afferent thin fibres.

The auditory pathway has been described using different methods of study throughout history: cell damage and degeneration, intracellular

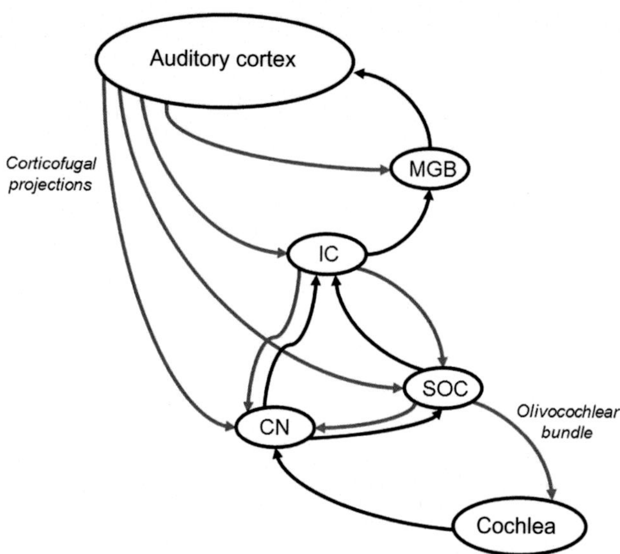

Figure 1.2 Schematic view of the ascending and descending auditory systems. *MGB*, medial geniculate body; *IC*, inferior colliculus; *CN*, cochlear nucleus; *SOC*, superior olivary complex. Arrows indicate the ascending and descending paths. *Modified from Terreros, G., Delano, P. 2016. Corticofugal modulation of peripheral auditory responses. Programa de Fisiología y Biofísica, ICBM, Facultad de Medicina, Universidad de Chile, Departamento de Otorrinolaringología, Hospital Clínico de, Santiago, Chile (Terreros and Delano, 2016).*

dyeing with tracers, deoxyglucose and by electrophysiological recording methods. By placing recording electrodes in various central nuclei bioelectrical responses, changes in the membrane potentials can be obtained from the auditory neurons that form the basis of evoked potentials measurable with a gross electrode. Evoked potentials, recorded in cats, shown in Fig. 1.3A, are examples of the averaged responses to a brief (click) sound stimuli. The differences between their shapes and, mainly, their latencies carefully reproduce the anatomical pathway, because activity evoked by a stimulus, first activates the receptors followed by the auditory nerve fibres, and subsequently the central nervous system (CNS), orderly ascending from nucleus to nucleus. Fig. 1.3B shows the short latency far-field potentials (brain-stem waves I–V). Recordings with a different time scale (50 ms) reveal the middle latency waves corresponding to the thalamic and cortical responses. A 500 ms time scale shows the full response including the late cortical potentials.

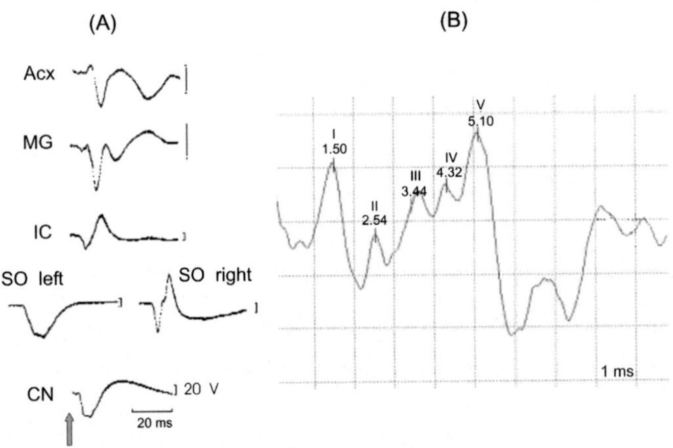

Figure 1.3 (A) Auditory local-field evoked potentials from a cat; and (B) human far-field evoked responses. An increasing latency is observed as the recording electrode is located higher in the pathway. The arrow shows the time of the auditory stimulus (a click). The human far-field recording shows from I (auditory nerve) to V (~IC) the normal latencies in milliseconds and the waves that roughly reflect the auditory nuclei in the brain stem.

AUDITORY NERVE EVOKED ACTIVITY

Beginning with the incoming sound, Table 1.1 exhibits the main mechanobioelectrical steps towards evoking an auditory nerve action potential.

The auditory nerve compound action potential (cAP) can be recorded from an electrode placed at the round window. A cAP averaged action potential is depicted in Fig. 1.4, left, with the two classical negative waves, N1 and N2, in response to clicks, that is, a stimulus that synchronizes the discharge of many nerve fibres. It reveals the activity of a group of single fibres and its synchronized discharges. The N1 amplitude is a function of the stimulus intensity as well as the number of synchronized fibres.

An auditory nerve single-fibre recording is shown in Fig. 1.4, right. A microelectrode may record the single-fibre activity when a stimulus of sufficient intensity is delivered. Its response can by characterized by a point of maximum sensitivity, that is, the response at the stimulus frequency with the lowest intensity, the characteristic frequency. The complete tuning curve includes the range of a fibre's threshold discharges over a range of stimulation frequencies.

Table 1.1 Events towards the generation of an auditory nerve action potential

Sound waves move the tympanic membrane
▼
The tympanic membrane moves the middle ear ossicles
▼
The ossicles move the oval window membrane
▼
The oval window movements produce motion of the cochlear fluids and basilar membrane
▼
The cochlear fluids and basilar membrane motion bend the inner hair cells' cilia
▼
The ciliary movements determine the excitation of the hair cells
▼
Finally, action potentials are generated at the auditory nerve fibres

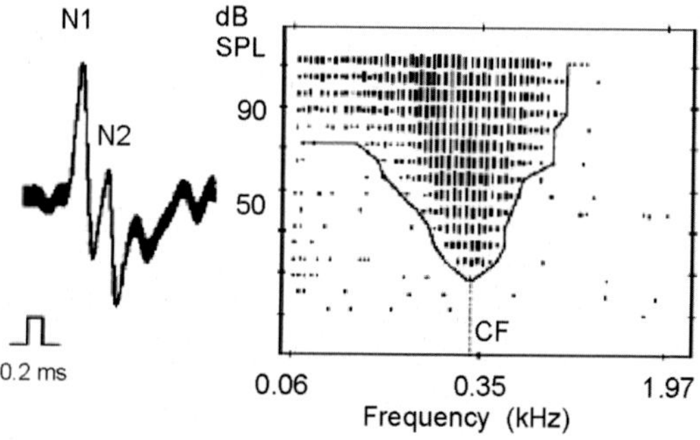

Figure 1.4 Guinea pig auditory nerve averaged compound action potential recorded from the round window during wakefulness (left) and the tuning curve of one auditory nerve fibre recorded with a microelectrode (right). *CF*, characteristic frequency.

EVOKED ACTIVITY

The different nuclei in the auditory pathway exhibit evoked potentials with characteristic waveforms and latencies according to their position in the brain. They are the first approach to how the brain *loci* may process information. The signal amplitude and the anatomical location of the recoding macroelectrode give us a first indication about the processing. Fig. 1.3A shows a series of local field recorded from a cat with electrodes placed in the auditory nuclei and cortex during wakefulness. The lower trace corresponds to the CN activation with the shorter latency, while the upper trace shows the primary auditory-cortical response exhibiting the longer latency. The two evoked potentials recorded from the right and left lateral olivary nuclei present a different waveform depending on the ear being stimulated. This is the lowest auditory nucleus in which the system may be acquainted with the side from which the sound is coming, indicating the initiation of the auditory binaural signals analysis.

Evoked resistance shifts associated with the evoked potentials in anaesthetized and awake cats were also recorded in parallel, demonstrating their close functional relationship. Time locked to the evoked potential, the evoked resistance shift consists of a brief resistance drop followed by a more prolonged rise recorded at all auditory nuclei but exhibiting a different wave shape and timing relative to the evoked potentials (Galambos and Velluti, 1968).

The far-field evoked potentials, mainly recorded in humans with scalp electrodes relative to mastoid electrodes, reveal a set of waves that correspond to an invading wave of negativity when sound stimuli are presented; that is, they are the responses of the negative activity throughout the brain stem beginning with the first wave that corresponds to the auditory nerve (I) activation.

A new human approach shows the influences of sleep and temperature on the far-field evoked potentials. Kräuchi et al. (2006) reported that proximal as well as distal increases in skin temperature are present during nap time. These reports are indicative of a great difference in temperature control during both night and nap sleep. Then, we began to study the brain-stem auditory evoked potential (BAEP) during nap sleep in human subjects. Volunteers with no pathology were stimulated with alternating clicks (80 dB SPL, 10/s) and BAEP were recorded together with online sleep polysomnographic control, under a hypnotic dose of chloral hydrate (CH). The experimental design was to study responses evoked by the stimuli during the afternoon nap (from 1:00 to 4:00 p.m.) with monitored skin temperature, and they showed a significant increase in wave V latency not related to body temperature but to sleep N2 (Fig. 1.5). Waves I–V (10 ms) also studied in the frequency domain (FFT) showed significant decreases in the power spectra of 400, 500 and 600 Hz during N2 sleep (Pedemonte et al., 2016).

COCHLEAR NUCLEUS

After entering the medulla, the auditory nerve fibres divide into two branches, called *anterior* and *posterior branches*. The classical division of the CN is shown in Fig. 1.6, in which the anterior branches innervate the

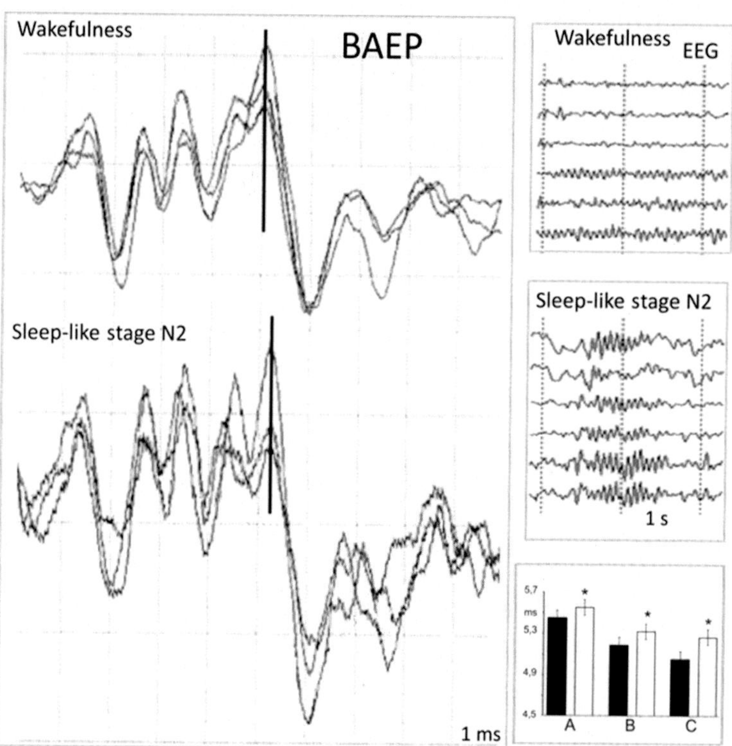

Figure 1.5 The brain-stem auditory potential (BAEP) was analysed during naplike sleep in human subjects. Volunteers with no pathology were stimulated with alternating clicks (80 dB SPL, 10/s) and BAEPs were recorded together with online sleep polysomnographic control, following a hypnotic dose of CH. The experimental design was to study the evoked responses during the afternoon nap (from 1:00 to 4:00 p.m.) with monitored skin temperature, and they showed a significant increase in wave V latency not related to body temperature but to sleep N2. *From Sensory processing in sleep: an approach from animal to human data. In: P. Perumal (Ed.), Synopsis of Sleep Medicine. APP, CRS Press (Chapter 22), pp. 379–396.*

ventral CN, while the posterior ones end in the posterior and dorsal CN (Lorente de Nó, 1981). The tonotopic organization coming from the cochlea is totally maintained in the CN as well as all throughout the auditory pathway.

The neurons located in each subnucleus exhibit diverse bioelectrical properties producing different excitatory neurotransmitters, such as glutamate and aspartate, and γ-aminobutiric acid (GABA) and glycine as inhibitory ones (Wenthold and Martin, 1984).

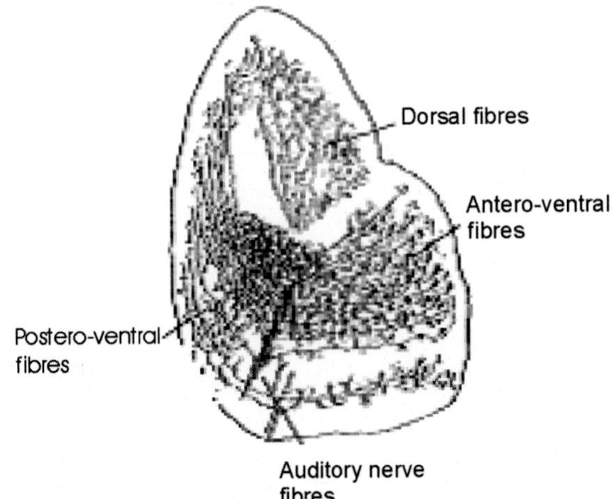

Figure 1.6 Cochlear nucleus neuroanatomical approach. *From Lorente de Nó, R. 1981 The Primary Acoustic Nuclei. Raven Press, New York.*

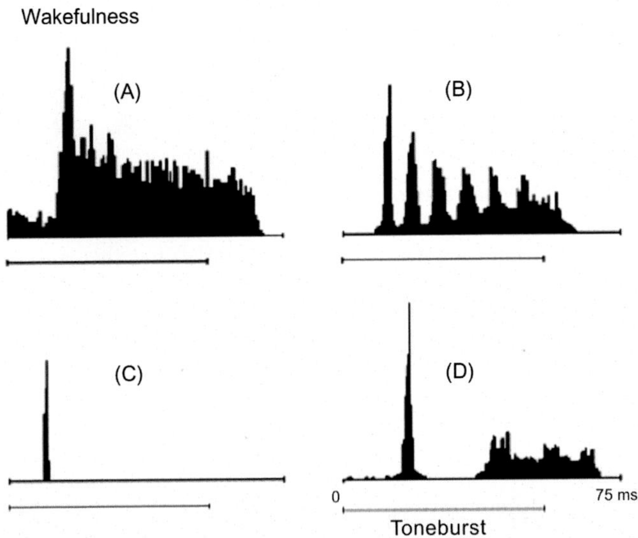

Figure 1.7 Cochlear nucleus neurons poststimulus time histogram (PSTH) classification: (A) primarylike, similar to the auditory nerve fibres discharges; (B) chopper response; (C) on response; (D) pauser, response to sound with a pause.

The electrophysiological classification of the CN unitary firing is based on discharge patterns (Pfeiffer, 1966). Fig. 1.7 shows examples of them: (A) primarylike, (B) onset, (C) chopper and (D) pauser. The firing pattern is not fixed, for each neuronal type but rather is a function of stimulus intensity and the animal behavioural state. Furthermore, the CN exhibits processes of excitation—inhibition of diverse origin, some intrinsic, from the CN itself, while others may arrive from higher centres, such as the IC, and even via direct projections from the auditory cortex (Schofield and Coomes, 2005).

The CN unitary neuronal activity, as well as at any other CNS region, is a representation of one or more networks' activity related to an auditory unit whose firing is always changing depending on several possibilities, such as stimulation characteristics or brain momentary behaviour.

The guinea pig is an excellent experimental model to develop auditory system research; it has a cochlea easy to manipulate as well as to control the sound stimulation applied directly in the middle ear. Furthermore, it is a friendly and peaceful animal that allows the electrophysiological recordings in a physiological way (without drugs). Almost 50% of the CN units increased their firing rate during slow wave sleep (SWS) and modified also in paradoxical sleep, showing an enhanced firing (Velluti et al., 1990).

Binaural fusion and localization

Listening to the same sound delivered to both ears, humans perceive a single phantom source between the ears or sometimes straight ahead in space. When sound to one ear is delayed, the source appears to shift to the other side. The clarity and compactness of perceived sources depend on the degree of similarity between the two signals. Addition of random noises to the signals can systematically vary the degree of binaural similarity. Such experiments showed that humans could perceive phantom sources even when the correlation of signals between the ears was as low as 0.3 (Jeffress et al.). Humans are not the only species that is so good at detecting binaural correlation in signals. Barn owls are known for their ability to locate prey by ear. Owls rapidly turn their heads toward sound sources, even when signals are delivered by earphones. This simple observation means that owls also perceive a single phantom source from binaural signals. Thus, tests similar to those used for humans could be applied to owls. The results were very similar to those obtained in humans. Owls could localize signals with binaural correlations as low as 0.3. The standard deviation of the mean localization errors remained almost constant from correlation 1 to correlation 0.3, below which the standard deviation increased sharply (Saberi et al., Konishi). The study with owls further advanced our knowledge about the neural mechanisms underlying binaural fusion and localization. The barn owl's brainstem auditory system consists of two pathways, one for processing the interaural time

(*Continued*)

(Continued)
difference (ITD) and the other for the interaural intensity difference (IID). These pathways converge in the external nucleus of the midbrain (inferior colliculus) to form a map of auditory space. The map contains space-specific neurons, which respond selectively to sounds coming from specific directions in space or specific combinations of ITD and IID. These neurons do not respond to uncorrelated pairs of sounds. They can detect small amounts (0.3) of correlation between signals going to the two ears. The responses of these neurons to partially uncorrelated signals can account for the behavioral results mentioned above.

Masakazu Konishi

Caltech Pasadena, California, USA (2008)

SUPERIOR OLIVARY COMPLEX

Three main nuclei are part of the superior olivary complex (SOC): the superior lateral, medial and trapezoid body nuclei. All of them exhibit tonotopy; that is, the tonal distribution established at the cochlea is maintained. A group of scattered nuclei surrounding the olive, called *periolivary*, is the origin of the olivocochlear efferent fibres. Three pathways provide the peripheral input to the olive and the lateral lemniscus nuclei: the ventral, medial and dorsal stria arising from the CN.

Local-field evoked potentials exhibit a clear difference when stimulated ipsi- or contralateral to the recording electrode position (Fig. 1.3A). Most of the lateral superior olive (LSO) neurons are excited by ipsilateral sound stimulation (E) while inhibited by contralateral sounds (I), called *EI units* (Fig. 1.8). Space-specific neurons in the owl's auditory map of the space involve computations that determine the interaural time (ITD) and interaural level (ILD) that define the auditory space. The neurons studied behaved like analog AND gates of ITD and ILD, suggesting the two inputs are multiplied instead of being added (Peña and Konishi, 2001). The same neurotransmitters found in the CN are also present in the superior olive complex. The binaural hearing process begins precisely here, at this complex. The main cues the brain uses to localize acoustic information from the environment are interaural intensity and interaural timing differences.

In our approach, some neurons in LSO lost the binaural capability during SWS, which will be shown in Chapter 5, Auditory Unit Activity in Sleep.

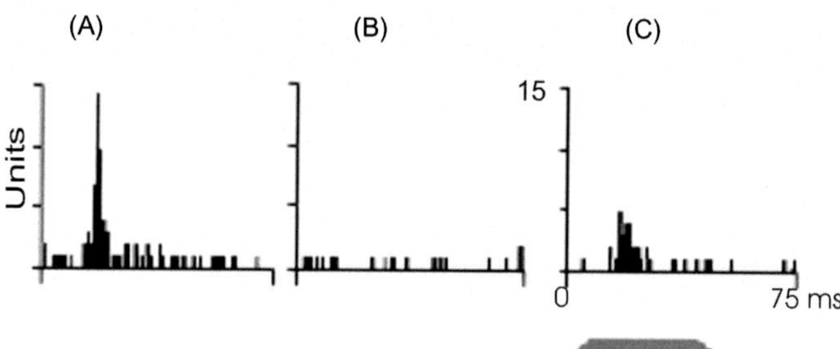

Figure 1.8 Awake guinea pig. PSTHs of a LSO neuron. (A) Ipsilateral sound stimulation showing a high probability of discharge at the beginning of the stimulus; (B) contralateral ear stimulation produces no response; (C) the binaural sound stimulation provokes a lower discharge in comparison to the ipsilateral one. Tone burst, 50 ms.

Neurotrasmitters, anatomy and function of the nuclei of the lateral lemniscus

The lateral lemniscus is a tract located in the most lateral region of the pons formed by the confluence of axons that originate from the neurons of the cochlear nuclei (CN), the superior olivary complex (SOC) and the nuclei of the lateral lemniscus (NLL), which project toward the inferior colliculus (IC) (Adams, 1997; Benson & Cant, 2008; Glendenning et al., 1981; Huffman & Covey, 1995; Kelly, van Adel, & Ito, 2009; Merchán & Berbel, 1996; Merchán, et al., 1994; Schofield & Cant, 1997; and Woollard & Harpman, 1940). The neurons of the NLL are interspersed among the bundles of fibres of the LL, and receive projections from the cochlear nuclei (large spherical and globular bushy, octopus, and type I multi-polar neurons) (Cant & Benson, 2003) and from the SOC (MNTB, LSO, and SPON) (Jack B Kelly et al., 2009). In view of their projections and their cytoarchitecture in most non-echolocating mammals, the NLL are organised into a dorsal subdivision (DNLL) that projects bilaterally, and another ventral one (VNLL) that projects unilaterally to the IC (Beyerl, 1978) (J C Adams, 1979). From an antomical point of view in the VNLL they can be differentiated a ventral and an intermediate (INLL) subdivisions (Glendenning, Brunso-Bechtold, Thompson, & Masterton, 1981b). However for other authors VNLL should be considered as a single nucleus from both an anatomical and functional perspective (Malmierca et al., 1998; Merchán & Berbel, 1996) (Riquelme et al., 2001) (Jack B Kelly et al., 2009). Furthermore, some authors identify an anterior division of the VNLL (Schofield & Cant, 1997). Also after electrophysiological studies of the responses of the neurons of the VNLL (R Batra & Fitzpatrick, 1997) recognize a medial division whose neurons reveal specific response patterns to temporal disparities (binaural

(*Continued*)

(Continued)

response neurons). In Bats have been found to have a specialised region related to the detection of temporal sound sequences linked to echolocation (Kutscher & Covey, 2009), called the columnar nucleus (VNLLc) located between the INLL and the VNLL (Vater et al., 1997).

Regarding the NLL, a concentric laminar pattern has been reported in the DNLL (Merchán et al., 1994) (Bajo, Merchán, et al. (1999), Bajo et al., 1993), with another one involving laminae that run parallel to the main axis of the lemniscus in the VNLL (Malmierca et al., 1998; Merchán & Berbel, 1996). As in the case of the fibro-dendritic laminae of the IC (Malmierca et al., 2005; Oliver et al., 1995; Oliver, 1987) (Loftus et al., 2010), these anatomical and functional differences may be the result of a complex, albeit unique, processing system for sound analysis.

Overall, the neurons of the NLL are largely GABA, and constitute one of the more powerful inhibitory inputs for the IC (Adams & Mugnaini, 1984. (Thompson et al., 1985, Moore & Moore, 1987; Roberts & Ribak, 1987; Winer et al., 1995; Zhang et al., 1998) (González-Hernández et al., 1996). However, the ascendant projection of the VNLL also uses the inhibitory neurotransmitter glycine, which besides regulating the processing of temporal sound properties is related to low-frequency suppression phenomena and, therefore, to across-frequency integration mechanisms (Peterson et al., 2009) (Yavuzoglu et al., 2010). Furthermore, dynorphin- and CRF-immunoreactive neurons that are immunonegative for GABA have been described in the most dorsal regions of the VNLL (INLL) (Ueyama et al., 1999).

Studies on the colocalization of GABA and glycine in the VNLL have revealed the existence of two immunocytochemical neuronal types in rats: non-GABA/non-glycine neurons of a possibly excitatory nature that prevail in the most dorsal region, and GABA/glycine neurons of an inhibitory nature that are more abundant in the most ventral region (Riquelme et al., 2001). In morphological terms, identification is made in the VNLL of globular bushy neurons and stellate neurons (J C Adams, 1979; Glendenning et al., 1981; Schofield & Cant, 1997). In addition to the globular bushy neuron type, subsequent studies (Zhao & Wu, 2001) have also identified three types of stellate neurons (I, II, and elongate). After in vivo intracellular recordings and neurobiotin injections in the most ventral third of the VNLL, a onset response has been demonstrate for globular bushy neuron (Nayagam et al., 2006). Besides its possible role in the spectral processing of sound, the refining of the coding of the temporal properties is most certainly the most important part the VNLL plays, involving an input convergence phenomenon and a sophisticated system for regulating inhibition. A key excitatory entry input for temporal processing comes from the octopus neurons of the cochlear nuclei, whose dendrites integrate the

(*Continued*)

(Continued)

signal from the primary afferents of different frequencies generating broadly tuned frequency responses. The special membrane properties of this neuron type (low voltage activated potassium (LVK) and hyperpolarisation activated cation (HCN) channels) permit accurate temporal analysis of sound stimuli (Golding et al., 1995; Oertel, 1999; Trussell, 1999). The globular bushy neurons of the most ventral region of the VNLL co-localize GABA and glycine and receive the calyces of the axons of the octopus neurons (Nayagam et al., 2005). The potentially excitatory neurons of the most dorsal regions of the VNLL vertical laminae receive excitatory inputs from the cochlear nuclei, and inhibitory ones from the ipsilateral MNTB and periolivary regions, in particular LNTB in most species (Covey & Casseday, 1986; Friauf & Ostwald, 1988; Glendenning et al., 1981b). The processing of the temporal information of the auditory stimulus in the VNLL includes not only monaural responses but also binaural ones for detecting interaural time disparities for the location of sound in space (R Batra & Fitzpatrick, 1997) (Ranjan Batra & Fitzpatrick, 2002). This function requires not only an extremely precise input from the cochlear nuclei (octopus neurons) but also a delicate balance between excitation and inhibition (Yang & Pollak, 1994a). From an anatomical perspective, the vertical laminae of the VNLL generate an inhibitory GABA/glycine input in the most ventral region, and another excitatory one in the more dorsal areas, whose convergence upon the IC leads to a refining of the responses to the location of the sound for the thalamus and cortex.

Experiments performed on brain slices involving intracellular recordings and the stimulation of lemniscal fibres have advanced our understanding of the response of the different functional types of neurons (onset, pause, adapting, regular and bursting) (Wu, 1999; Zhao & Wu, 2001) of the VNLL and its specific pharmacological blocking of postsynaptic receptors (Irfan et al., 2005). In particular this last paper have confirmed the existence of AMPA receptors (which block the early/short EPSPs after stimulating the lemniscus). Also subunits of the AMPA receptor, in particular GluR4 for imparting fast gating and desensitization kinetics to the receptor complex, have been described in the neurons of the VNLL (Caicedo & Eybalin, 1999; Petralia & Wenthold, 1992; Sato et al., 2000). Elsewhere, in experiments conducted by Irfan et al., 2005, the IPSPs are blocked by strychnine and/or by bicuculline, which implies the existence of receptors for GABAA and glycine. Prior studies on the binding of receptors have provided evidence of specificity for strychnine and the alpha 1 glycine subunit. The concurrence of GABAA and glycine receptors in the neurons of the VNLL may be due to projections from nuclei whose neurons use these neurotransmitters, such as MNTB (glycine) or VNTB or LNTB (GABA), as well as to the reception of inputs from the inhibitory neurons of the VNLL,

(*Continued*)

(Continued)

which co-localize both neurotransmitters (Riquelme et al., 2001b). The potential effect of the co-release of GABA and glycine on the precision of the temporal codes has yet to be fully understood (Noh et al., 2010), although it is accepted that glycine is a neurotransmitter that refines the inhibitory response and allows very high temporal resolutions in the detection of coincidence of inputs (Brand et al., 2002; Pecka et al., 2008).

The DNLL has a topographic organization defined by concentric laminae arranged oppositely to the VNLL in an orthogonal direction to the longitudinal axis of the LL (Bajo et al., 1999; Henkel, 1997; Merchán et al., 1994). The DNLL is made up of GABA neurons that projects bilaterally to the IC (Joe C Adams & Mugnaini, 1984). It is considered a specialised structure for processing binaural information, more specifically of interaural intensity differences (IID) (Li & Kelly, 1992). The neurons of the DNLL persistently inhibit auditory stimuli, with the effect being the refining of the responses for the location of sound, and in particular in bats it mediates echo suppression (J B Kelly & Kidd, 2000; Yang & Pollak, 1994b) (Burger & Pollak, 2001; Pecka et al., 2007) (Faingold, Anderson, & Randall, 1993). Its neurons synthesized GABA (Adams & Mugnaini, 1984), and its connections mediate in the IC the release and recapture of this inhibitory neurotrasmitter (Shneiderman et al., 1993). Neuropharmacological studies following blocking with NBQX and CNQX in DNLL neurons have revealed the involvement of NMDA receptors in the long-lasting component and AMPA in the early component of the synaptic response (J B Kelly & Kidd, 2000) (Porres, Meyer, Grothe, & Felmy, 2011). Ultrastructural assays have revealed the GABAergic, inhibitory nature of the crossed connection via Probst's commissure, and confirm the existence of a topographically organised system of a reciprocal nature in the DNLL (Oliver & Shneiderman, 1989), which is no doubt that is relevant for explaining phenomena of persistent inhibition in the electrophysiological responses of its neurons (Pecka et al., 2007).

Miguel Merchan
Universidad de Salamanca, Spain

INFERIOR COLLICULUS

The key anatomical location of the inferior collicus (IC) is suitable for interactions among different possible connections, that is, the lower auditory nuclei and cortex, reticular formation, periaqueductal grey, superior colliculus, and the like, as well as among diverse physiological states such as the sleep-wakefulness cycle. A role of the IC may be to serve as a synaptic crossroad for ascending and descending information, a true auditory processing-integrative function that also depends on the general state of the CNS.

Three subnuclei constitute its internal structure: the central, external and dorsal nuclei (Huffman and Henson, 1990; Oliver and Shneiderman, 1991). The central one is a main station for all ascending information, organized in layers associated with a tonotopic distribution, with low frequencies at dorsal *loci* while higher frequencies are located in ventral regions (Oliver and Morest, 1984).

Animal recordings of IC local-field potentials and far-field potential recordings in humans exhibit some characteristics corresponding to the local multiunitary activity and the negative wave converging on the IC region, respectively. Neuron unitary firing recordings may be classified using the temporal distribution over the poststimulus time histograms (Morales-Cobas et al., 1995; Velluti 1997) (Fig. 1.9).

Intracellular, in vivo recordings of physiologically identified IC central nucleus auditory neurons and their subthreshold membrane potential activities were recorded (Pedemonte et al., 1997). The spontaneous action potentials were divided into two groups according to their duration and mean firing rate. Current injection revealed adaptation and membrane potential changes outlasting the electrical stimuli by 20–30 ms. Sequences of synaptic potentials, longer than the sound, were observed lasting up to 90 ms with binaural stimulation. The data from Pedemonte et al. (1997) are consistent with the existence of a multisynaptic pathway by which signals arrive at the IC central nucleus, including corticofugal

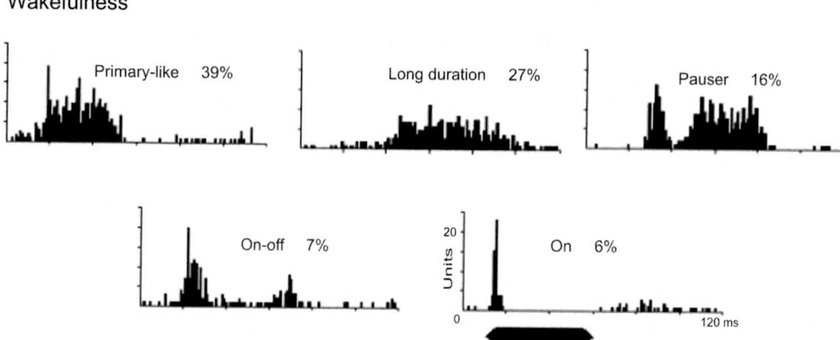

Figure 1.9 PSTHs from inferior colliculus units in awake guinea pigs. The recordings, carried out with glass micropipettes and no drugs, demonstrate five discharge patterns: (A) primarylike, (B) long lasting, (C) with notch, (D) on-off and (E) on.

pathways. This may contribute to the long duration postsynaptic potentials.

The IC excitatory neurotransmitters are amino acids and the inhibitory ones are GABA and to a lesser extent glycine (Faingold et al., 1991). In addition the IC receives noradrenergic terminals from the *locus coeruleus* cells.

In anaesthetized guinea pigs, there is little spontaneous activity, whereas during wakefulness all IC neurons analysed showed a high rate of spontaneous firing and a great number of membrane potential oscillations (Torterolo et al., 1995).

Most IC cells are binaurally sensitive (Caird, 1991). The binaural cues are also the ITDs and the IIDs. The characteristic firing of such units led to the creation of a space map in the IC equivalent of the barn owl (Knudsen and Konishi, 1978). This means that particular neurons may respond to stimuli coming from particular regions of the surrounding space, thus creating an auditory map of the world.

In addition, the frequency tuning properties of the IC neurons did not show any significant change during the cortex cooling period. The results demonstrate that AC cooling inactivates excitatory corticofugal pathways and results in a less activated intrinsic inhibitory network in the IC (Popelar et al., 2016).

MEDIAL GENICULATE BODY

This thalamic nucleus is divided into three regions: ventral, dorsal and medial. The afferents to the medial geniculate (MG) body come from the IC and the reticular formation, while its efferent fibres reach the auditory cortex. The ventral portion exhibits a laminar structure and maintains a net tonotopic organization. Each cell of the lamina receives input from the IC, which also receives descending fibres from cortical origin (Winer, 1991).

The unitary activity of the MG is similar to that of the IC with inhibitory lateral bands. A great proportion of the cells in the MG are binaural, responding to the ITDs while others are predominantly sensitive to the IIDs. Axons from the thalamic reticular nucleus cells, all of which are GABAergic, also terminate in the ventral nucleus (Montero, 1983) while some sparse cholinergic immunoreactive axons occur in the same nucleus with unknown source (Levey et al., 1987). A specific auditory reticular nucleus is located in the reticular thalamus contributing to the information transfer to the cortex (Guillery et al., 1998) as it will be shown under reticular formation.

AUDITORY CORTEX

The human auditory cortex (AI) lies in the superior temporal plane; Heschl's gyrus exhibits a Nissl staining revealing a seven layer organization originally shown by Ramón y Cajal (1952; Fig. 1.10).

The tonotopy is also exhibited by the cortex. A general view of cortical physiology is a tonotopic core surrounded by a belt of tissue with a less clear tonotopic representation and flanked by areas that show no evidence of tonotopy at all (Fig. 1.11). The direction of the tonotopic gradient in AI is defined caudomedially for high frequencies while low

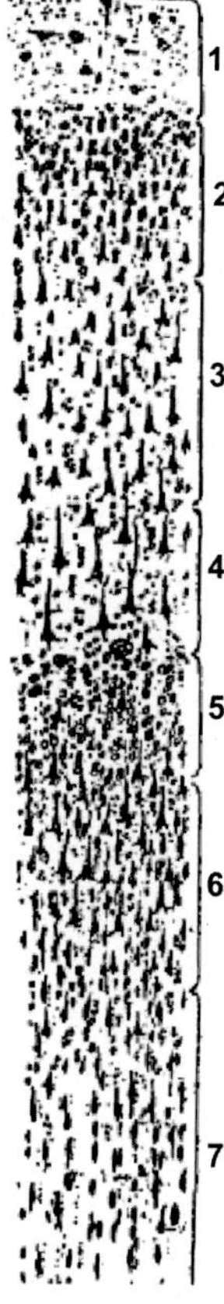

Figure 1.10 Nissl staining histological study by Ramón y Cajal described seven laminae in the human auditory cortex. Plexiform layer (1), two granular laminae (2, 5) and the 3 and 4 laminae showing a different pyramidal neuron sizes. Other authors join six or seven laminae into only a single one, lamina 6. *Modified from Histologie du systéme nerveux de l'homme & des vertébrés, Ramón y Cajal (1909).*

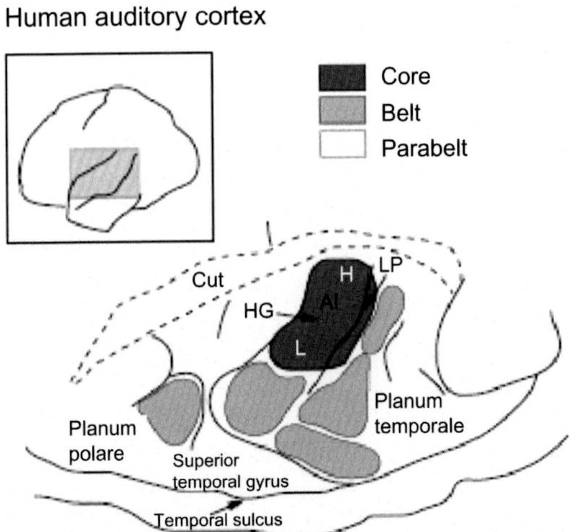

Figure 1.11 In humans the auditory cortices lie on the superior temporal plane. HG, Heschl gyrus; LP, lateroposterior area; AI, auditory primary area; H, high frequency; and L, low frequency. *Modified from Semple, M.N., Scott, B.H. 2003. Cortical mechanisms in hearing. Curr. Opin. Neurobiol. 13: 167–173.*

frequencies are found rostrolaterally (Semple and Scott, 2003). Neurons located in the MG body provide input to all auditory cortices. The thalamocortical projections are the basis for a particular sequential processing through the core, belt and parabelt auditory areas. Then, auditory signals may be distributed to parietal, temporal and frontal multimodal areas.

The 'what/where' distinction has gained support from animal and human studies (Semple and Scott, 2003). The 'where', spatial processing, implicates the dorsal pathways linking the caudomedial belt to frontal and parietal targets. The 'what', objects, phonemes, originates in the anterior core and belt areas reaching targets in the temporal lobe. Human imaging studies (Zatorre et al., 2002) are more or less consistent with the 'what/where' hypothesis, while a significant correlation between temporal and parietal cortical responses have revealed the possibility of important cross talk between the systems (Alain et al., 2001).

Adding to the notion that the auditory system depends on the brain state, the cat cortical local-field potential recording exhibits a long latency response whose amplitude is dependent on the state of the experimental animal, such as anaesthetized or awake.

Plasticity of behavioural responses to sounds can be observed when the behavioural state of an animal changes (e.g., from sleep to waking, from nonattentive to attentive; Ehret, 1997). When averaged across neurons, sound-evoked activity in two auditory-cortical areas was surprisingly well preserved during sleep. Neural responses to acoustic stimulation were present during both slow-wave and rapid-eye movement sleep, were repeatedly observed over multiple sleep cycles, and demonstrated similar discharge patterns to the responses recorded during wakefulness in the same neuron. Our results suggest that the thalamus is not as effective a gate for the flow of sensory information as previously thought. At the cortical stage, a novel pattern of activation/deactivation appears across neurons (Peña et al., 1999; Issa and Wang, 2008).

'Top-down' inputs during selective attention reshape the auditory-cortex neuronal receptive fields to help filter relevant stimulus features.

The short-term plasticity caused by selective attention is restricted to non-primary auditory-cortical areas. Most of the earlier fMRI studies have simply probed whether significant hemodynamic response enhancements can be seen in nonprimary and primary auditory-cortical responses to sounds when they are selectively attended versus ignored (Jääskeläinen and Ahveninen, 2014).

When memories are covertly cued via auditory or olfactory stimulation, investigations of these subtle manipulations of memory processing during sleep can help elucidate the mechanisms of memory preservation in the human brain (Oudiette and Paller, 2013).

UNITARY ACTIVITY

The neuronal activity recorded in the primary cortical area (AI) during W in guinea pigs and monkeys, in response to pure tone stimulation (Fig. 1.12) represents an approach to the central processing of sound by each auditory neuron integrated in a neural network(s). The different poststimulus time histograms (PSTHs) shown, that is, the probability of discharge over time, are not fixed for each neuron but change their characteristics depending on several factors, such as the stimulus intensity, the stimulus spectrum (e.g., noise or tone bursts), CNS awake or asleep (Recanzone, 2000).

In our experimental results with guinea pigs during wakefulness and sleep, we have never encountered a bursting neuron. A burst is a group of action potentials with 4 ms interspike interval preceded by a silent period of about 100 ms. It was reported that auditory thalamus or cortical neuron bursts occur in synchronized electroencephalogram states, SWS in a low proportion, or under anaesthesia. During wakefulness some bursting

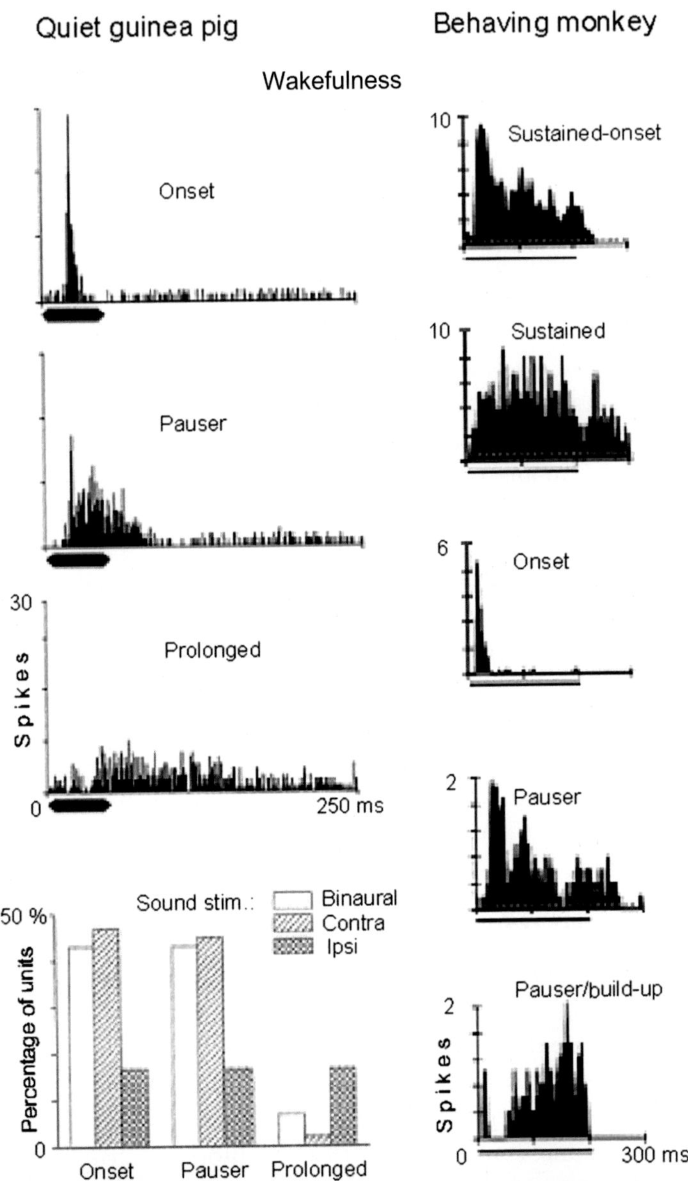

Figure 1.12 Left, neuronal response patterns in the AI auditory cortex of the awake guinea pig. PSTHs of the three main response types obtained for contralateral best frequency 50-ms tone bursts stimuli during wakefulness. Bottom, response type distribution for monaural ipsicontralateral and binaural acoustic stimulation. Right, responses from neurons of the behaving macaque monkey classified into five categories. Each PSTH shows eight trials near the characteristic frequency of different AI neuron. The differences may be due to species differences, animal condition such as quiet wakefulness versus behaving, stimulation time, the method of stimulus delivery, closed versus open field, and the like. *Modified from Peña, J.L., Pérez-Perera, L., Bouvier, M., Velluti, R.A. 1999. Sleep and wakefulness modulation of the neuronal firing in the auditory cortex of the guinea-pig. Brain Res., 816: 463–470 and Recanzone, G. 2000. Response profiles of auditory cortical neurons to tones and noise in behaving macaque monkeys. Hear. Res. 150: 104–118.*

neuron may appear, although in a very low frequency being the greatest presence under anaesthesia (Massaux and Edeline, 2003; Edeline, 2005).

Tuning Curves

Similar to the auditory nerve curve shown in Fig. 1.4, right, several tuning curves have been described from the auditory cortex. Single-peaked tuning curves are narrow, excitatory and V shaped, in response to pure tone stimuli. Others are scarcely tuned, exhibiting wide curves or two peaked and wide curves, without exhibiting a net characteristic frequency (Schreiner et al., 2000).

Magnetoencephalography

Magnetoencephalography (MEG) is characterized by its very high temporal resolution, on the order of milliseconds, as compared with neuroimaging techniques (temporal resolution on the order of minutes). MEG has theoretical advantages over EEG for detecting cortical dipole localization, because there is less effect from current conductivity caused by cerebrospinal fluid, skin, and so on, while the spatial resolution for MEG evoked activity is on the order of a few millimetres. Thus, the dipole generated may be detected with high precision. During wakefulness a multiwave response to pure tone stimuli was observed in the human primary auditory cortex with waves at 50, 100, 150 and 200 ms (Kakigi et al., 2003; see Chapter 4, Auditory Information Processing During Sleep).

THE HIPPOCAMPAL THETA RHYTHM-AUDITORY UNIT RELATIONSHIP

Phase locking between auditory neuron firing and hippocampal theta rhythm (Hipp θ) during wakefulness at different pathway levels, such as colliculus and primary cortex, has been demonstrated (Pedemonte et al., 1996; 2001; Pedemonte and Velluti, 2005; Velluti et al., 2000; Velluti and Pedemonte, 2002, 2012). We postulated that the Hipp θ could play a role as an internal clock adding a temporal dimension to the auditory processing at auditory pathway and cortical *loci* (see Chapter 5, Auditory Unit Activity in Sleep).

IMAGING

Functional imaging of cerebral cortical activity relies on the coupling of blood flow to neuronal firing and metabolism with functional magnetic resonance (fMRI). The relationship between the hemodynamic fMRI-based signals and neuronal firing is a matter of discussion. Mukamel et al. (2005) compared the human auditory-cortical single-unit activity (two recorded patients) with the fMRI of 11 healthy people analysed in response to identical auditory stimuli. A highly significant correlation between single-unit activity and the fMRI results was obtained, meaning the fMRI may provide reliable reflection of the human auditory neuronal firing.

The extent of fMRI activation in the superior temporal gyrus increases with stimulus intensity, revealing that intensity lower than 60 dB SPL would make activation impossible to detect. The stimulus frequency also has an important impact: fMRI activation is greater with a stimulus at 1 kHz and stepped tones than with single pure tones. Several stimuli have been used, such as tones, words, and so forth. Music is the stimulus evoking the largest activation of the primary and secondary as well as associative cortical areas (Fig. 1.13).

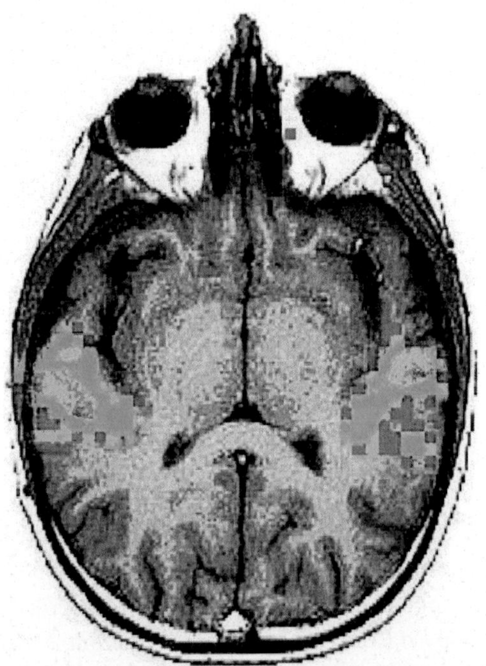

Figure 1.13 Auditory functional MR imaging with music. The intensities of brain regions activation varied from high (yellow) to low (blue). The primary auditory cortex is bilaterally activated while the activity of secondary auditory cortex is greater on the left. *Modified from Bernal, B., Altman, N. 2001. Auditory functional MR imaging. Am. J. Roentgenol. 176: 1009–1015 (Bernal and Altman, 2001).*

Neurotransmitters

The principal cortical neurotransmitter is GABA, known from the cortical supragranular zone in AI. The use of iontophoretic GABA injections demonstrated an inhibition on neurons responding to sound. Noradrenaline also has been described as an inhibitor, although with a slower time-course than GABA. Cholinergic cortical neurons inhibit about half of the neurons while the other half is facilitated by acetylcholine. The cholinergic actions on AI have a long latency and a long duration of several minutes. Some of these actions are blocked by atropine.

The analysis of neurotransmitter roles in the auditory pathway, particularly in the IC in sleep and wakefulness, can also contribute to the understanding of neural information processing. The results reported address the afferent and efferent actions in which excitatory neurotransmission is involved. With respect to the NMDA action on IC cells, it can be concluded that there are no differences at the cellular level between sleep and waking (Goldstein-Daruech et al., 2002). Therefore, other types of excitatory neurotransmission, as well as synaptic inhibition, may underlie the physiologically complex changes that occur in the sleeping brain. The excitatory efferent system was postulated as being mainly glutamatergic (Feliciano and Potashner, 1995). Therefore, the effect of auditory (AI) electrical stimulation was mimicked, a few minutes later, by iontophoretic kynurenic acid ejection onto the same IC neuron, supporting the notion of an efferent pathway acting through excitatory amino acid receptors on IC cells or on inhibitory neurons (Goldstein-Daruech et al., 2002; Velluti and Pedemonte, 2002). Thus, the descending auditory system can participate in the processing of the incoming information. Excitatory amino acid transmission would be present in sleep and wakefulness, acting by similar mechanisms together with several other neurotransmitters on the same neuron.

THE RETICULAR FORMATION

Also receiving sensory information through collateral fibres of the sensory pathways in general and the auditory pathway in particular (Huttenlocher, 1960), the reticular formation is an important structure because of the general awakening role that may it play (Moruzzi and Magoun, 1949). From the functional point of view, this region is part of a diffuse spreading of information as well as orderly information flow

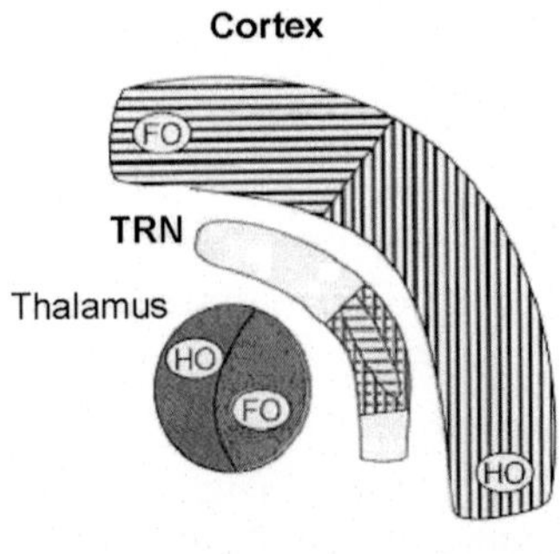

Figure 1.14 Thalamic reticular nucleus (TRN). Three reticular layers can be recognized in the auditory region of the TRN. The central TRN layer receives stimuli from FO, called *first-order nuclei,* and their main input from ascending specific fibres and the cortical horizontal stripes. Inner and outer layers receive input from higher order thalamic and cortical regions, vertical stripes. HO, called *higher order nuclei,* receive their main driving input from cortical layer V. *Modified from Guillery et al. (1998).*

toward the superior processing centres. Moreover, specific sensory nuclei are located in the thalamic reticular formation, including a well-defined reticular auditory nucleus; this particular auditory reticular region is connected to other thalamic regions and to the auditory cortex (Fig. 1.14).

THE CEREBELLUM

The cerebellum constitutes another auditory processing centre whose function is still under research. The auditory information reaches the cerebellum as evidenced by evoked potentials recorded, for example, at the rat vermis (Lorenzo et al., 1977) and the evoked resistance shifts in cats (Galambos and Velluti, 1968). In coordination with the visual and somesthetic neuronal activity of the superior colliculus, this cerebellar auditory information may underlie functions such as the control of head movements during the search of a sound source in space and perhaps also in learning. Cerebellar vermis electrical stimulation introduces changes in the cochlear nerve and microphonic activity, thus contributing to efferent system functions (Velluti and Crispino, 1979).

Then, it has been established that the evoked potentials, the magnetoencephalographic evoked activity and the unitary neuronal firing (recorded at different loci of the auditory pathway) exhibit changes related to different physiological brain states, for instance, the sleep-wakefulness cycle, the habituation process and learning. Finally it can be assumed that all the changes mentioned are due to the interactions with other auditory system components, such as the complex and powerful efferent system.

THE EFFERENT DESCENDING SYSTEM

The CNS uses are several mechanisms to control or modulate the incoming auditory signals. I mention the following examples: (1) ear movement (mainly in lower mammals), (2) movement of the middle ear muscles, (3) regulation of the mechanical properties of the outer and inner hair cell of the organ of Corti, (4) actions over the auditory nerve primary afferents fibres (5) and actions on each nucleus of the auditory pathway.

Moreover, Hernández-Peón et al. (1956) found a reduction in the evoked potentials recorded from the CN in cats while receiving stimuli of other sensory modalities. This pioneer study suggested the presence of corticofugal effects, which later showed up in other experimental works.

How is the Efferent System Organized?

There are two opinions about this subject. One supports the notion of three interconnected feedback circuits: Cortex-thalamus-mesencephalon; mesencephalon-SOC-CN; and SOC-CN-cochlea. The second position supports the idea of a continuous descending chain with actions and feedback loops at all levels (Spangler and Warr, 1991). Important experimental data support is the second position: the possibility of producing similar effects to those that activate the olivocochlear bundle through stimuli in the higher regions of the auditory system (Desmedt, 1975). Due to the complexity of the system, it becomes difficult to support a single position, and because both possibilities can coexist (Fig. 1.15).

Moreover, signals originating in the cerebellar vermis have been reported to modify the auditory cochlear responses, thus contributing to the central control of this sensory input at the receptor level (Velluti and Crispino, 1979). The periaqueductal grey has also been shown to have an influence on (as well as an anatomical connection to) the CN unitary activity (Pedemonte et al., 1990; Radmilovich et al., 1991).

AUDITORY CORTEX EFFERENTS

Two outgoing cortical systems are described: one connected to the MG nucleus and the other, with a wider distribution, is projected toward the IC, the nonauditory thalamic nuclei, the *striatum* and the lateral pontine region (Spangler and Warr, 1991).

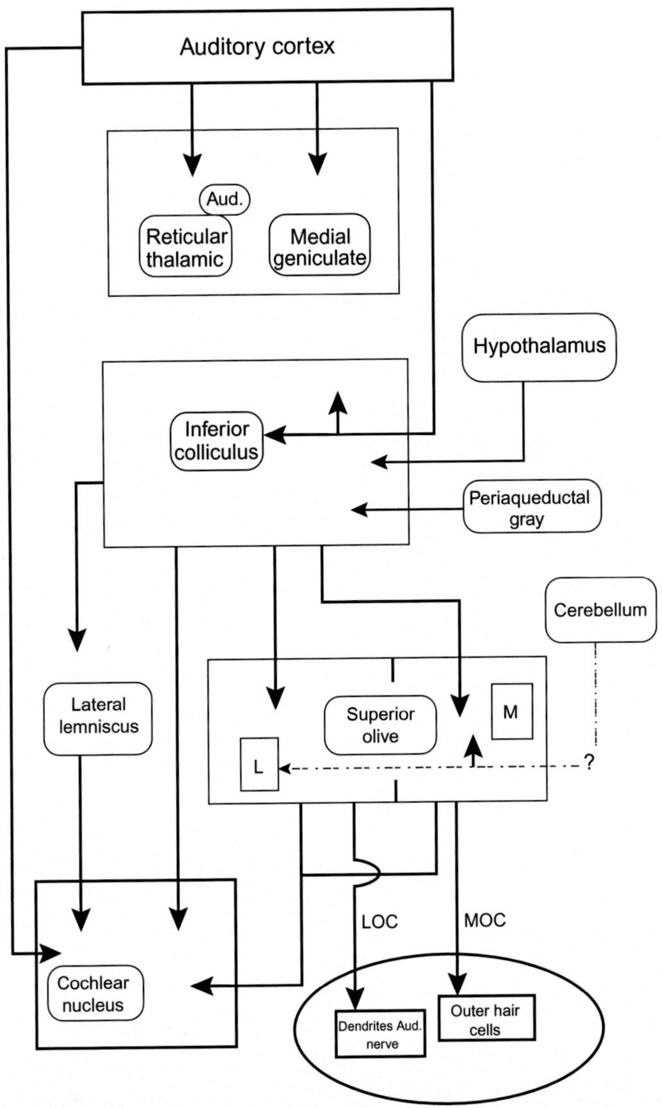

Figure 1.15 General diagram of the efferent (descending) pathways. L, lateral superior olive; M, medial superior olive; LOC, lateral olivocochlear fibres; MOC, medial olivocochlear fibres.

Auditory corticofugal fibres constitute a descending parallel pathway (Winer, 2006) reaching the MG monosynaptically (Winer, Diehl, and Larue, 2001), the IC (Winer et al., 1998), the CN (Schofield and Coomes, 2005) and the olivocochlear neurons (Mulders and Robertson, 2000). The axons descending from the cortex to the IC terminate in the different regions of this nucleus, including the central one (Faye-Lund, 1986; Saldaña et al., 1996; Saldaña and Merchán, 1992), showing the descending fibres in the IC innervate wide zones that go beyond the specifically auditory central nucleus and the contralateral IC.

Scarce data have been published about the effects of the auditory cortex on the neuronal activity of the MG nucleus. Actions of the electrically stimulated primary auditory cortex on the MG have been reported by Sauerland et al. (1972). The authors showed the existence of presynaptic excitability increase in this nucleus, a phenomenon that is capable of modulating the auditory input to the cortex.

On the other hand, electrical microstimulation of the auditory cortex by means of a chronic intracerebral multielectrode array produced a significant reduction of otoacustic emission, while there was no change under stimulation of nonauditory cortical areas.

Little is known about the physiological aspects of the dorsal efferent system. Both excitatory and inhibitory potentials in neurons of the IC in response to ipsilateral electric stimulation of the auditory cortex have been reported (Mitani et al., 1983). The same was demonstrated for the central and dorsal nucleus of the IC with extracellular recordings (Syka and Popelar, 1984; Sun et al., 1989). The existence of bilateral cortical actions on the activity of the same IC auditory neuron has been described (Torterolo et al., 1998). These results suggest the possibility of more complex processes that would include the integration of the ascending information plus the efferent actions of both cortices on the same collicular neuron. Moreover, firing shifts may be obtained by stimulating the AI as well as other cortical regions, such as the orbital gyrus (Sun et al., 1989; Torterolo et al., 1998; Sauerland et al., 1972).

This excitatory descending system was postulated as being mainly glutamatergic (Feliciano and Potashner, 1995), whereas cortical electrical stimulation affects the ICc neurons, decreasing the firing of 86% of the recorded neurons and increasing the firing in the remaining 14% (Fig. 1.16). Approximately the same proportions were obtained when, with a double micropipette, a NMDA blocker was electrophoretically injected on the same ICc unit (Goldstein-Daruech et al., 2002).

INFERIOR COLLICULUS EFFERENTS

This nucleus has multiple connections to the cortex, which provides a high capacity parallel channel for processing auditory information. Evidence for extensive and reciprocal hypothalamic and IC connections

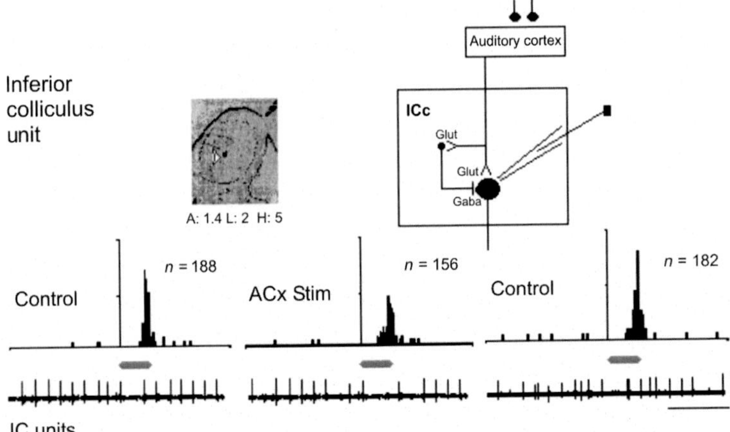

Figure 1.16 Effect of auditory cortex (AI) on an auditory inferior colliculus neuron. PSTH show that cortical electrical stimulation produces a decrement in the response to sound and the possible GABAergic influence. *ICc*, inferior colliculus central nucleus; *ACx stim*, auditory cortex stimulation; *Glut*, glutamate; *GABA*, gamma amino butyric acid; *n*, number of spikes; sound stimuli: contralateral tone burst, 50 ms duration at the unit characteristic frequency; 70 dB SPL. *Modified from Goldstein-Daruech et al. (2002).*

have been reported (Winer, 2006). From the IC, descending fibres project to the superior olive and the CN (Huffman and Henson, 1990; Saldaña et al., 1996). The projection to the superior olive ends in the neurons that originate in the olivocochlear system, so serving as connection from higher centres to the most peripheral regions of the efferent system (Faye-Lund, 1986).

THE OLIVOCOCHLEAR SYSTEM

The olivocochlear efferent neurons originate in the brain stem and terminate in the organ of Corti, thereby allowing the CNS to control cochlear and auditory nerve function. This bundle of fibres, first observed by Lorente de Nó (1933) in connection with the CN, has been extensively described by Rasmussen (1946, 1960). Primarily, the efferent fibres split in one crossed bundle and one direct bundle, although nowadays it is agreed that there exists (1) a medial olivocochlear system (MOC) and (2) a lateral one (LOC) (Warr, 1975; Guinan, 1986). The medial system, whose neurons are located in the superior medial olive on either side, sends its axons to the outer hair cells. The lateral system, with its neurons located in the periphery of the LSO –, ipsi and contralateral make synapse with the afferent auditory fibres beneath the inner hair cells. Another important anatomical difference between the medial efferent fibres and the lateral ones is that the former are myelinated, while the latter are unmyelinated (Warr, 1992).

Acetylcholine (ACh) is the main neurotransmitter released by the MOC. LOC is also cholinergic although it is now suggested that LOC fibres colocalize many neurotransmitters and neuromodulators. It was proposed that the efferent innervation of outer hair cells is mediated by the heteromeric α9 and α10 subunits of the nicotinic ACh receptor (Elgoyhen et al., 2001).

Physiological Approaches

Galambos (1956) experimentally demonstrated the capacity of the olivary-cochlear bundle to reduce the amplitude or to block the cAP by electrically stimulating the direct and crossed bundles of efferent fibres at the floor of the IVth ventricle. The evoked decrease of the cAP is proportional to the intensity of the electric stimulation and the stimulating sound. Subsequently, Fex (1962, 1967), using single-unit recordings, reported that such effects corresponded to an inhibitory process and added the observation that the cochlear microphonic (CM) increased in amplitude in response to electric stimulation of the floor of the IVth ventricle.

On the other hand, after systemic injection of benzodiazepines, we reported an inverse change (Velluti and Pedemonte, 1986): While the cAP increased, the CM diminished its amplitude. Then, we postulated a central action of benzodiazepine on GABAergic neurons, probably onto efferent periolivary neurons, that would determine the actions on both, the receptor (CM) and the auditory nerve (cAP). The experimental control, the peripheral injection of the same benzodiazepine, did not have any effect when applied directly to the cochlea.

The conclusion is that all possibilities are present in the system; that is, the cAP and the CM amplitude may increase or decrease, depending on the way the system is activated. Different experimental approaches, exhibited in Table 1.2, produce all the possible changes in both potentials amplitudes.

Table 1.2 Research evolution of the cochlear recorded potentials

Authors	Experimental conditions	cAP shifts			CM shifts		
Galambos (1956)	Anaesthetized cats. Electrical stimulation on the floor of the IVth ventricle		►				
Fex (1962)	Anaesthetized cats. Electrical stimulation on the floor of the IVth ventricle					►	
Buño et al. (1966)	Awake behaving guinea pigs. Habituation to constant stimuli. Without ossicles and closed sound stimulating system	W	►	W	W	►	W
Oatman (1971)	Awake cats without middle ear ossicles cAP and changing attention with visual stimulation	W	►	W			
Velluti and Pedemonte (1986)	Awake guinea pigs with systemic injected benzodiazepine. Without ossicles and closed sound stimulating system	W	►	W	W	►	W

Velluti et al. (1989)	Awake and sleeping guinea pigs Without ossicles and closed sound stimulating system	W	►	SWS	W	►	SWS
Pedemonte et al. (2004), Pavez et al. (2006) and Velluti et al. (1989)	Awake and sleeping guinea pigs Without ossicles and closed sound stimulating system	W	►	SWS	W	►	SWS
Délano et al. (2007)	Awake chinchillas Cochlear recorded potentials and attention shifts	W	►	W	W	►	W

cAP, auditory nerve compound action potential; CM, cochlear microphonic; W, wakefulness; SWS, slow wave sleep.

Relationships with Attention Processes

Experimental approaches demonstrated that cochlear recorded potential amplitudes may also shift in a parallel fashion. Under different physiological conditions, such as habituation and dishabituation to sound (Buño et al., 1966) and during sleep (Velluti et al., 1989), the cAP and the CM amplitudes change in the same way, the two simultaneously recorded potentials increasing or decreasing. Accordingly, Oatman (1971, 1976) provides information correlating the activity of the efferent system to attention processes: When a laboratory animal is performing a visual task, the amplitude of the auditory nerve potential shows a decrement. Moreover, it has recently been reported that visuospatial attention modulates cochlear amplitudes, decreasing the cAP and concomitantly increasing the CM amplitudes (Délano et al., 2007; Table 1.2).

MODIFICATIONS OF THE MECHANICAL STATUS OF THE COCHLEA

Electromotility, the ability to produce movements such as contraction and stretching of the outer hair cells (Brownell et al., 1985) permits the control of the displacements of the basilar membrane, thus influencing the cochlear mechanics and, consequently, the transduction process and the sensitivity of the receptor system, such as the inner hair cells. This capacity to change the sensitivity of the system is called the *cochlear amplifier*. In accordance with the definition of Robles and Ruggero (2001), the 'cochlear amplifier' is a positive feedback process that increases the sensitivity to the response of the basilar membrane in response to low-intensity stimuli. Scarce experimental data are available concerning the effects of the efferent system on the mechanical control of the basilar membrane.

Otoacoustic Emissions

Acoustic emissions are detectable sounds in the external auditory canal generated by the cochlea (Kemp, 1978). A series of experiments have shown the influences of the medial efferent system on otoacoustic emissions, which provide indirect information about the motility of the basilar membrane. Distortion products are tones produced in the cochlea through distortion as a response to stimulation by two primary tones, f1 and f2, with different frequencies. The wave generated at the level of the distorting frequency, due to nonlinearities, which are probably located in the basilar membrane, expands in both directions from its point of origin at the point of maximum response to f2. Efferent stimulation may decrease the distortion product amplitudes or sometimes increase them (Mountain, 1980; Siegel and King, 1982). The effects of the medial efferent system on otoacoustic emissions seem to indicate a mechanical change in the cochlea. At low stimulus sound levels, depression of the basilar membrane movement by stimulating the medial efferent system appears as the dominant mechanism for modulating inner hair cell function. This changes the receptor sensitivity and, thus, can change the afferent fibre discharge rate. The mechanism that could underlie efferent system modification of the cochlear vibration would be the electromotility of the outer hair cells. However, it has been reported that an additional cellular mechanism, which concerns active movements of the cilia, could contribute to increased system sensitivity and could likely also be part of the origin of the otoacoustic emissions (Martin and Hudspeth, 1999).

GENERAL CONSIDERATIONS OF EFFERENT ACTIVITIES

The discrepancy between the natural environmental noise level and the experimental high-level noise used to evoke the medial olivocochlear bundle activity, evidence that the MOC system did not evolve to protect the ear from natural sound acoustic trauma (Kirk and Smith, 2003). Only in rare instances are ambient noise levels sustained at moderately high intensities, for example, the highest noise in a natural environment corresponds to frog choruses reaching about 90 dB SPL (Narins and Hurley, 1982). By contrast, all experiments in which a MOC-mediated protective effect was observed use a much higher sound intensity, ~100–150 dB SPL. In addition, the MOC is present in all mammals evolved over 170 million years in the total absence of sustained high-intensity and narrow frequency natural noise.

The global functions of the efferent system and particularly the olivocochlear partition are not well known. Hence, the main hypothesis for its function, are to

1. Improve the detection of a signal masked in noise. The MOC system very possibly evolved for unmasking significant sound-reducing simultaneous low-level noise, a hypothesis with wide experimental support.
2. Modify the mechanical status of the cochlea by providing control of the outer hair cells and cilia movements influencing inner hair cell activity and otoacoustic emissions.

3. Control some natural body-produced noises. High-intensity internal noise, such as the chewing, respiratory, heart and blood flow noises shall be controlled to maintain the system in optimal functional conditions. Other physiological noise, such as the circulation through the carotid arteries located close to the cochlea, is blocked and not consciously perceived although they are generating continuous acoustic input reflecting body physiology: the CN single-unit firing synchronized with heart beats-blood flow, reported by Velluti et al. (1994).
4. I now offer a new hypothesis regarding efferent fibre function. The efferent system, which is connected to a great diversity of auditory-related regions, exerts actions over the input and the processing at different levels, which means that efferent activity, can establish a synergy between the auditory system and the ever-changing CNS status. Synergy means to put both the brain status and the auditory input reciprocal interactions to work in order to support the widely distributed brain changes occurring on entering sleep. This hypothesis is introduced to explain the efferent actions of all sensory systems and the auditory system in particular. The continuous interaction/adaptation between the CNS and the outside and inside world (the body) can be achieved only through systems (receptors, afferent and efferent pathways) that are capable of constantly following both rapid and slow changes of the state of the CNS.

Finally, this new key function presented for the auditory efferent system is supported by several experimental approaches (Velluti, 1997, 2005; Velluti et al., 2000; Velluti and Pedemonte, 2002; Pedemonte and Velluti, 2005) particularly those coming from system physiology of both behaving, awake, or asleep animals and humans.

REFERENCES

Alain, C., Arnott, S.R., Hevenor, S., Graham, S., Grady, C.L., 2001. "What" and "where" in the human auditory system. Proc. Natl Acad. Sci. USA 98, 12301–12306.

Brownell, W.E., Bader, C.R., Bertrand, D., de Ribaupierre, Y., 1985. Evoked mechanical response of isolated cochlear outer hair cell. Science 277, 194–196.

Buño, W., Velluti, R., Handler, P., Garcia-Austt, E., 1966. Neural control of the cochlear input in the free guinea-pig. Physiol. Behav. 1, 23–35.

Caird, D., 1991. Processing in the colliculi. In: Altshuler, R.A., Bobbin, R.P., Clopton, B.M., Hoffman, D.W. (Eds.), Neurobiology of Hearing: Central Auditory System. Raven Press, New York, pp. 253–292.

Délano, P.H., Elgueda, D., Hamame, C.M., Robles, L., 2007. Selective attention to visual stimuli reduces cochlear sensitivity in chinchillas. J. Neurosci. 27, 4146–4153.

Desmedt, J.E, 1975. Physiological studies of the efferent recurrent auditory system. In: Keidel, W.D., Neff, D. (Eds.), Handbook of Sensory Physiology. Springer, Berlin, pp. 219–246.

Edeline, J.-M., 2005. Learning-induced plasticity in the thalamo-cortical auditory system: should we move from rate to temporal code descriptions? In: Konig, R., Heil, P., Budinger, E., Scheich, H. (Eds.), The Auditory Cortex. Lawrence Erlbaun Ass, Mahwah, NJ, London, pp. 365–382.

Ehret, G., 1997. The auditory cortex. J. Comp. Physiol. A 181, 547–557.

Elgoyhen, A.B., Vetter, D.E., Katz, E., Rothlin, C.V., Heinemann, S.F., Boulter, J., 2001. α10: a determinant of nicotinic cholinergic receptor function in mammalian vestibular and cochlear mechanosensory hair cells. Proc. Natl. Acad. Sci. U.S.A. 98, 3501–3506.

Faingold, C.L., Gehlbach, G., Caspary, D., 1991. Functional pharmacology of inferior colliculus neurons. In: Altshuler, R.A., Bobbin, R.P., Clopton, B.M., Hoffman, D.W. (Eds.), Neurobiology of Hearing: Central Auditory System. Raven Press, New York, pp. 223–251.

Faye-Lund, H., 1986. Projections from the inferior colliculus to the superior olivary complex in the albino rat. Anat. Embryol. 175, 35–52.

Feliciano, M., Potashner, S., 1995. Evidence for a glutamatergic pathway from the guinea-pig auditory cortex to the inferior colliculus. J. Neurochem. 65, 1348–1357.

Fex, J., 1962. Auditory activity in centrifugal and centripetal cochlear fibers in the cat. Acta Physiol. Scand. 55, 2–68.

Fex, J., 1967. Efferent inhibition in the cochlea related to hair-cell activity: a study of postsynaptic activity of the crossed olvo-cochlear fibers in the cat. J. Acoust. Soc. Am. 41, 666–675.

Galambos, R., 1956. Suppression of auditory nerve activity by stimulation of efferent fibers to cochlea. J. Neurophysiol. 19, 424–437.

Galambos, R., Velluti, R.A., 1968. Evoked resistance shifts in unanesthetized cats. Exp. Neurol. 22, 243–252.

Goldstein-Daruech, N., et al., 2002. Effects of excitatory amino acid antagonists on the activity of inferior colliculus neurons during sleep and wakefulness. Hear. Res. 168, 174–180.

Guillery, R.W., Feig, S.L., Lazsádi, D.A., 1998. Paying attention to the thalamic reticular nucleus. Trends Neurosci. 21, 28–32.

Guinan, J.J., 1986. Effects of efferent neural activity on cochlear mechanics. Scand. Audiol. Suppl. 25, 53–62.

Hernandez-Peon, R., Scherrer, H., Jouvet, M., 1956. Modification of electric activity in cochlear nucleus during attention in unanesthetized cats. Science. 123 (3191), 331–332.

Huffman, R.F., Henson, O.W., 1990. The descending auditory pathway and acousticomotor systems: connections with the inferior colliculus. Brain Res. Rev. 15, 295–323.

Huttenlocher, P.R., 1960. Effects of the state of arousal on click responses in the mesencephalic reticular formation. Electroencephgr. Clin. Neurophysiol. 12, 819–827.

Issa, E.B., Wang, X., 2008. Sensory responses during sleep in primate primary and secondary auditory cortex. J. Neurosci. 28 (53), 14467–14480.

Jääskeläinen, J., Ahveninen, J., 2014. *Review article*: Auditory-cortex short-term plasticity induced by selective attention auditory-cortex short-term plasticity induced by selective attention. Neural Plasticity 2014, 1–11.

Kakigi, R., Naka, D., Okusa, T., Wang, X., Inui, K., Qiu, Y., et al., 2003. Sensory perception during sleep in humans: a magnetoencephalograhic study. Sleep Med. 4, 493–507.

Kemp, D.T., 1978. Stimulated acoustic emissions from within the human auditory system. J. Acoust. Soc. Am. 64, 1386–1391.

Kirk, E., Smith, D.W., 2003. Protection from acoustic trauma is not a primary function of the medial olivocochlear efferent system. J. Assoc. Res. Otolaryngol. 4, 445–465.

Knudsen, E.I., Konishi, M., 1978. A neural map of auditory space in the owl. Science 4343, 795–797.

Kräuchi, K., Knoblauch, V., Wirz-Justice, A., Cajochen, C., 2006. Challenging the sleep homeostat does not influence the thermoregulatory system in men: evidence from a nap vs. sleep-deprivation study. Am. J. Physiol.- Regul. Integr. Comp. Physiol. 290, 1052–1061.

Levey, A.I., Hallanger, A.E., Wainer, B.H., 1987. Choline acetyltransferase immunoreactivity in the rat thalamus. J. Comp. Neurol. 257, 317–332.

Lorente de Nó, R., 1933. Anatomy of the eight nerve. III. General plan of structure of the primary cochlear nuclei. Laryngoscope 43, 327–350.

Lorente de Nó, R., 1981. The Primary Acoustic Nuclei. Raven Press, New York.

Lorenzo, D., Velluti, J.C., Crispino, L., Velluti, R.A., 1977. Cerebellar sensory functions: rat auditory evoked potentials. Exp. Neurol. 55, 629–636.

Massaux, A.E., Edeline, J.-M., 2003. Burst in the medial geniculate body: a comparison between anaesthetised and unanaesthetized guinea pig. Exp. Brain Res. 153, 573–578.

Martin, P., Hudspeth, A.J., 1999. Active hair bundle movements can amplify a hair cells's response to oscillatory mechanical stimuli. Proc. Natl. Acad. Sci. USA. 96, 14306–14311.

Mitani, A., Shimokouchi, M., Nomura, S., 1983. Effects of the stimulation of the primary auditory cortex upon colliculo-geniculate neurons in the inferior colliculus of the cat. Neurosci. Lett. 42, 185–189.

Moller, A.R., Rollins, P.R., 2002. The non-classical auditory pathways are involved in hearing in children but not in adults. Neurosci. Lett. 319, 41–44.

Montero, V.M., 1983. Ultrastructure identification of axon terminals from thalamic reticular nucleus in the medial geniculate body in the rat: an EM autoradiographic study. Exp. Brain Res. 51, 338–342.

Morales-Cobas, G., Ferreira, M.I., Velluti, R.A., 1995. Sleep and waking firing of inferior colliculus meurons in response to low frequency sound stimulation. J. Sleep Res. 4, 242–251.

Moruzzi, G., Magoun, H., 1949. Brain stem reticular formation and activation of the EEG. Electroencephgr. Clin. Neurophysiol. 1, 455–473.

Mountain, D.C., 1980. Changes in endolymphatic potential and crossed olivococlear bundle stimulation alter cochlear mechanics. Science 210, 71–72.

Mukamel, R., Gelbard, H., Arieli, A., Hasson, U., Fried, I., Malach, R., 2005. Coupling between neuronal firing, field potentials, and FMRI in human auditory cortex. Science 5736, 951–954.

Mulders, W.H.A.M., Robertson, D., 2000. Evidence for direct cortical innervation of medial olivocochlear neurons in rats. Hear. Res. 144, 65–72.

Narins, P.M., Hurley, D.D., 1982. The relationship between call intensity and function in the Puerto Rican Coqui (Anura: Leptodactylidae). Herpetologica 38, 287–295.

Oatman, L.C., 1971. Role of visual attention on auditory evoked potentials in unanesthetized cats. Exp. Neurol. 32, 341–356.

Oatman, L.C., 1976. Effects of visual attention on the intensity of auditory evoked potentials. Exp. Neurol. 51, 41–53.

Oliver, D.L., Morest, D.K., 1984. The central nucleus of the inferior colliculus of the cat. J. Comp. Neurol. 222, 237–264.

Oliver, D.L., Shneiderman, A., 1991. The anatomy of the inferior colliculus: a cellular basis for integration of monaural and binaural information. In: Altshuler, R.A., Bobbin, R.P., Clopton, B.M., Hoffman, D.W. (Eds.), Neurobiology of Hearing: Central Auditory System. Raven Press, New York, pp. 195–222.

Oudiette, D., Paller, K.A., 2013. Upgrading the sleeping brain with targeted memory reactivation. Trends Cogn. Sci. 17 (3), 142–149.

Pavez E., Drexler D., Délano P.H., Pedemonte M., Falconi A., Robles L., et al. (2006) Efectos del ruido y gentamicina sobre los potenciales cocleares en chinchilla y cobayo. XXII Congreso Latinoamericano y I Iberoamericano de Ciencias Fisiológicas. Buenos Aires, Argentina. Physiological Mini-Reviews 2 (4): 64.

Pedemonte, M., Drexler, D., Velluti, R.A., 2004. Cochlear microphonic amplitude reduction and variability after noise exposure and gentamicin. Hear. Res. 194, 25–30.

Pedemonte, M., Velluti, R.A., 2005. Sleep hippocampal thetarhythm and sensory processing. In: Lander, M., Cardinali, D.P., Perumal, P. (Eds.), Sleep and Sleep Disorders: A Neuropsychopharmacological Approach. Landes Biosciencies. TX/Springer, NY, pp. 8–12.

Pedemonte, M., Peña, J.L., Velluti, R.A., 1990. Periaqueductal gray influence on anteroventral cochlear nucleus unitary activity and naloxone effects. Hear. Res. 47, 219–228.

Pedemonte, M., Peña, J.L., Torterolo, P., Velluti, R.A., 1996. Auditory deprivation modifies sleep in the guinea-pig. Neurosci. Lett. 223, 1–4.

Pedemonte, M., Torterolo, P., Velluti, R.A., 1997. *In vivo* intracellular characteristics of inferior colliculus neurons in guinea pigs. Brain Res. 759, 24–31.

Pedemonte, M., Pérez-Perera, L., Peña, J.L., Velluti, R.A., 2001. Sleep and wakefulness auditory processing: cortical units vs. hippocampal theta rhythm. Sleep Res. 4, 51–57.

Pedemonte, M., Medina-Ferret, E., Velluti, R.A., 2016. Sensory processing in sleep: an approach from animal to human data. In: Perumal, P. (Ed.), Synopsis of Sleep Medicine. (APP, CRS) Academic Press., (CRS) Taylor and Frances Press, Boca Raton, pp. 379–396. Chapter 22.

Peña, J.L., Konishi, M., 2001. Auditory spatial receptive fields created by multiplication. Science 292, 179–189.

Peña, J.L., Pérez-Perera, L., Bouvier, M., Velluti, R.A., 1999. Sleep and wakefulness modulation of the neuronal firing in the auditory cortex of the guinea-pig. Brain Res. 816, 463–470.

Pfeiffer, R.R., 1966. Anteroventral cochlear nucleus: waveforms of extracellularly recorded spike potentials. Science 154, 667–668.

Popelar, J., Suta, D., Lindovský, J., Bures, Z., Pysanenko, K., Chumak, T., et al., 2016. Cooling of the auditory cortex modifies neuronal activity in the inferior colliculus in rats. Hear. Res. 332, 7–16.

Radmilovich, M., Bertolotto, C., Peña, J.L., Pedemonte, M., Velluti, R.A., 1991. A search for e mesencephalic pariaqueductal gray-cochlear nucleus connection. Acta Physiol. Pharmacol. Latinoam. 41, 369–375.

Ramón y Cajal, S., 1952. Histologie du Système Nerveux. Consejo Superior de Investigaciones Científi cas. Madrid, España.

Rasmussen, G.L., 1946. The olivary peduncle and other fiber projections of the superior olivary complex. J. Comp. Neurol. 84, 141–219.

Rasmussen, G.L., 1960. Efferent fibers of cochlear nerve and cochlear nucleus. In: Rasmussen, G.L., Windle, W.F. (Eds.), Neural Mechanisms of the Auditory and Vestibular Systems. Thomas, Springfield, IL, pp. 105–115.

Recanzone, G., 2000. Response profiles of auditory cortical neurons to tones and noise in behaving macaque monkeys. Hear. Res. 150, 104–118.

Robles, L., Ruggero, M.A., 2001. Mechanics of the mammalian cochlea. Physiol. Rev. 81, 1305–1352.

Saldaña, E., Merchán, M.A., 1992. Intrinsic and commissural connections of the rat inferior colliculus. J. Comp. Neurol. 319, 417–437.

Saldaña, E., Feliciano, M., Mugnaini, E., 1996. Distribution of descending projections from primary auditory neocortex to inferior colliculus mimics the topography of intracollicular projections. J. Comp. Neurol. 371, 15–40.

Sauerland, E., Velluti, R., Harper, R., 1972. Cortically induced changes of presynaptic excitability in higher-order auditory afferents. Exp. Neurol. 36, 79–87.
Schofield, B.R., Coomes, D.L., 2005. Auditory cortical projections to the cochlear nucleus in guinea pigs. Hear. Res. 199, 89–102.
Schreiner, C.E., Read, H.L., Sutter, M.L., 2000. Modular organization of frequency integration in primary auditory cortex. Annu. Rev. Neurosci. 23, 501–509.
Semple, M.N., Scott, B.H., 2003. Cortical mechanisms in hearing. Curr. Opin. Neurobiol. 13, 167–173.
Siegel, J.H., King, D.O., 1982. Efferent neural control of cochlear mechanics? Olivococlear bundle stimulation affects cochlear biomechanical nonlinearity. Hear. Res. 6, 171–182.
Spangler, K.M., Warr, W.B., 1991. The descending auditory system. In: Altshuler, R.A., Bobbin, R.P., Clopton, B.M. (Eds.), Neurobiology of Hearing. The Central Auditory System. Plenum Press, New York, pp. 27–45.
Syka, J., Popelar, J., 1984. Inferior colliculus in the rat: neuronal responses to stimulation of the auditory cortex. Neurosci. Lett. 31, 235–240.
Sun, X., Jen, P.H.S., Sun, D., Zhang, G., 1989. Corticofugal influences on the responses of bat inferior collicular neurons to sound stimulation. Brain Res 495 (1), 1–8.
Torterolo, P., Pedemonte, M., Velluti, R.A., 1995. Intracellular in vivo recording of inferior colliculus auditory neurons from awake guinea-pigs. Arch. Ital. Biol. 134, 57–64.
Torterolo, P., Zurita, P., Pedemonte, M., Velluti, R.A., 1998. Auditory cortical efferent actions upon inferior colliculus unitary activity. Neurosci. Lett. 249, 172–176.
Velluti, R.A., 1997. Interactions between sleep and sensory physiology. A review. J. Sleep Res. 6, 61–77.
Velluti, R.A., 2005. Remarks on sensory neurophysiological mechanisms participating in active sleep processes. In: Parmeggiani, P.L., Velluti, R.A. (Eds.), The Physiologic Nature of Sleep. Imperial College Press, London, pp. 247–265.
Velluti, R.A., Crispino, L., 1979. Cerebellar actions on cochlear microphonics and on auditory nerve action potential. Brain Res. Bull. 4, 621–624.
Velluti, R.A., Pedemonte, M., 1986. Differential effects of benzodiazepines on choclear and auditory nerve responses. Electroencephalogr. Clin. Neurophysiol. 64, 556–562.
Velluti, R.A., Pedemonte, M., 2002. In vivo approach to the cellular mechanisms for sensory processing in sleep and wakefulness. Cell. Mol. Neurobiol. 22, 501–515.
Velluti, R.A., Pedemonte, M., 2012. Sensory neurophysiologic functions participating in active sleep processes. Sleep Sci. 5 (4), 103–106.
Velluti, R., Pedemonte, M., García-Austt, E., 1989. Correlative changes of auditory nerve and microphonic potentials throughout sleep. Hear. Res. 39, 203–208.
Velluti, R.A., Pedemonte, M., Peña, J.L., 1990. Auditory Brain Stem Unit During Sleep Phases. Pontenagel Press, Bochum, pp. 94–96.
Velluti, R.A., Pena, J.L., Pedemonte, M., Narins, P.M., 1994. Internally generated sound stimulate cochlear nucleus units. Hear. Res. 72, 19–22.
Velluti, R.A., Pedemonte, M., Peña, J.L., 2000. Reciprocal actions between sensory signals and sleep. Biol. Signals Recept. 9, 297–308.
Warr, W.B., 1975. Olivocochlear and vestibular efferent neurons of the feline brain stem. Their location, morphology, and number determined by retrograde axonal transport and acetylcholinesterase histochemistry. J. Compar. Neurol. 161, 159–182.
Warr, W.B., 1992. Organization o f olivococlear efferent system in mammals. In: Webster, D.B., Popper, A.N., Fay, R.R. (Eds.), Mammalian Auditory Pathway: Neuroanatomy. Springer-Verlag, New York, pp. 410–448.
Wenthold, R.J., Martin, M.R., 1984. Neurotransmitters of the auditory nerve and central auditory system. In: Berlin, C. (Ed.), Hearing Science: Recent Advances. College-Hill Press, San Diego, CA, pp. 341–369.

Winer, J.A., 1991. Anatomy of the medial geniculate body. In: Altshuler, R.A., Bobbin, R.P., Clopton, B.M., Hoffman, D.W. (Eds.), Neurobiology of Hearing: Central Auditory System. Raven Press, New York, pp. 293–333.

Winer, J.A., 2006. Decoding the auditory corticofugal systems. Hear. Res. 212, 1–8.

Winer, J.A., Diehl, J.J., Larue, D.T., 2001. Projections of auditory cortex to the medial geniculate body of the cat. J. Comp. Neurol. 430, 27–55.

Zatorre, R.J., Bouffard, M., Ahad, P., Belin, P., 2002. Where is "where" in the human auditory cortex? Nat. Neurosci. 5, 906–909.

FURTHER READING

Cutrera, R., Pedemonte, M., Vanini, G., Goldstein, N., Savorini, D., Cardinali, D.P., et al., 2000. Auditory deprivation modifies biological rhythms in golden hamster. Arch. Ital. Biol. 138, 285–293.

Demanez, J.P., Demanez, L., 2003. Anatomophysiology of the central auditory nervous system: basic concepts. Acta Oto-Rhino-Laryngol Belg. 57, 227–236.

Edeline, J.-M., 2003. The thalamo-cortical auditory receptive fields: regulation by the sates of vigilance, learning and neuromodulatory systems. Exp. Brain Res. 153, 554–572.

Edeline, J.-M., Manuta, Y., Hennevin, E., 2000. Auditory thalamus neurons during sleep: changes in frequency selectivity, threshold and receptive field size. J. Neurophysiol. 84, 934–952.

Liberman, M.C., 1990. Effects of chronic cochlear de-efferentation on auditory-nerve response. Hear. Res. 49, 209–224.

Pedemonte, M., Peña, J.L., Morales-Cobas, G., Velluti, R.A., 1994. Effects of sleep on the responses of single cells in the lateral superior olive. Arch. Ital. Biol. 132, 165–178.

Peña, J.L., Pedemonte, M., Ribeiro, M.F., Velluti, R.A., 1992. Single unit activity in the guinea-pig cochlear nucleus during sleep and wakefulness. Arch. Ital. Biol. 130, 179–189.

Winer, J.A., Larue, D.T., Diehl, J.J., Hefti, B.J., 1998. Auditory cortical projections to the cat inferior colliculus. J. Comp. Neurol. 400, 147–174.

CHAPTER 2

The Physiological Bases of Sleep

All living creatures, being plants or animals, unicellular or complexly organized, oscillate in time-configuring rhythms, which arise as a result of sensory modulation acting on genetically encoded information. By using the electroencephalography (EEG), it is possible to typify each behavioural state: wakefulness (W), stage I and II, slow wave sleep (SWS) with stages N1 and N3, and paradoxical sleep (PS) with or with no rapid eye movement (REM), whose overnight sequence shows a characteristic architecture that constitutes an ultradian rhythm. The physiology of many different systems is modulated by the wakefulness–sleep cycle: The processing of sensory information, the oneiric activity, the cardiovascular and respiratory functions, the endocrine functions, as well as body temperature control, homoeostasis and the energetic metabolism, all happen to change depending on the moment of the cycle. A number of neural centres and networks are involved in the generation and maintenance of this cycle. Much has been speculated about the possible action(s) of sleep, but little is known about this state essential for life, which takes a third of it.

The concept of cell assembly has been one of the bases for the development of mathematical models of distributed memories, an enterprise performed by various researchers around 1970. These models showed how information encoded in thousands of parallel firing neurons can be stored into a large neuronal network. The mathematical representation of the global activity of a large cell assembly is a large dimensional numerical vector. This vector is composed by numbers that measure the biophysical activities of the neurons (e.g., firing rates). Some memory models store pairs of input–output vectors (Mizraji, 2008; unpublished data; Velluti and Pedemonte, 2012).

Moreover, the efforts of many years have converged on a new hypothesis about the functions of sleep (the synaptic homoeostasis hypothesis), which claims that that sleep maintains synaptic homoeostasis. Summarizing, sleep is the price we have to pay for plasticity, and its function would be the homoeostatic regulation of the total synaptic weight impinging on neurons.

The Auditory System in Sleep
DOI: https://doi.org/10.1016/B978-0-12-810476-7.00002-6

During sleep, neurons in the cerebral cortex fire and stop firing together in waves of activity having frequencies of less than 4.5 Hz. Such slow-wave activity, whose most pronounced EEG feature is nonrapid eye movement (NREM) sleep, is also a reliable predictor of sleep intensity (Tononi and Cirelli, 2003, 2005).

Although, electrophysiological sensory data, particularly auditory data, keep reaching cortical areas during both NREM sleep and PS or REM sleep (Peña et al., 1999; Issa and Wang, 2008; Velluti and Pedemonte, 2012).

SLEEP IS A DIFFERENT CENTRAL NERVOUS SYSTEM STATE

Throughout evolution, more and more complex systems have arisen to control the narrow margins of normality. However, a normal range is not constant but presents oscillations and rhythms.

Although it is possible to recognize rhythms of activity and rest in practically all living creatures, sleep, with all its particular physiological characteristics, appears in homoeotherm vertebrates. Throughout phylogenetic evolution this rhythm is found to be ultradian in most of species (Esteban et al., 2005). Even though the whole central nervous system (CNS) participates in this state, certain entities, such as the basal forebrain, are related mainly to the SWS organization, while the *dorsolateralis pontine tegmentum* contributes the characteristic expressions or signs of PS. Sleep arises as an ensemble of physiological changes, where different systems take part under the regulation of the CNS, particularly the 85% of the whole system, the cortical mantle (Fig. 2.1).

Polysomnography

Studies of EEG rhythms carried out in humans, allow us to classify four stages of sleep (Loomis et al., 1938), being PS first described as a separate stage by Aserinski and Kleitman (1953). The polysomnogram is a continuous and simultaneous recording of physiological variables during sleep. The minimum recording for stage identification includes EEG, electromyogram (EMG), and electro-oculogram (EOG). The particular combinations of the three bioelectrical signals – amplitude and frequency of the waves in the EEG, electromyographic and oculomotor activity – permit the recognition of the different sleep stages. Thus both wakefulness with its variations and sleep with its stages (SWS or orthodox or non-REM sleep and PS or active sleep or REM sleep) can be defined using

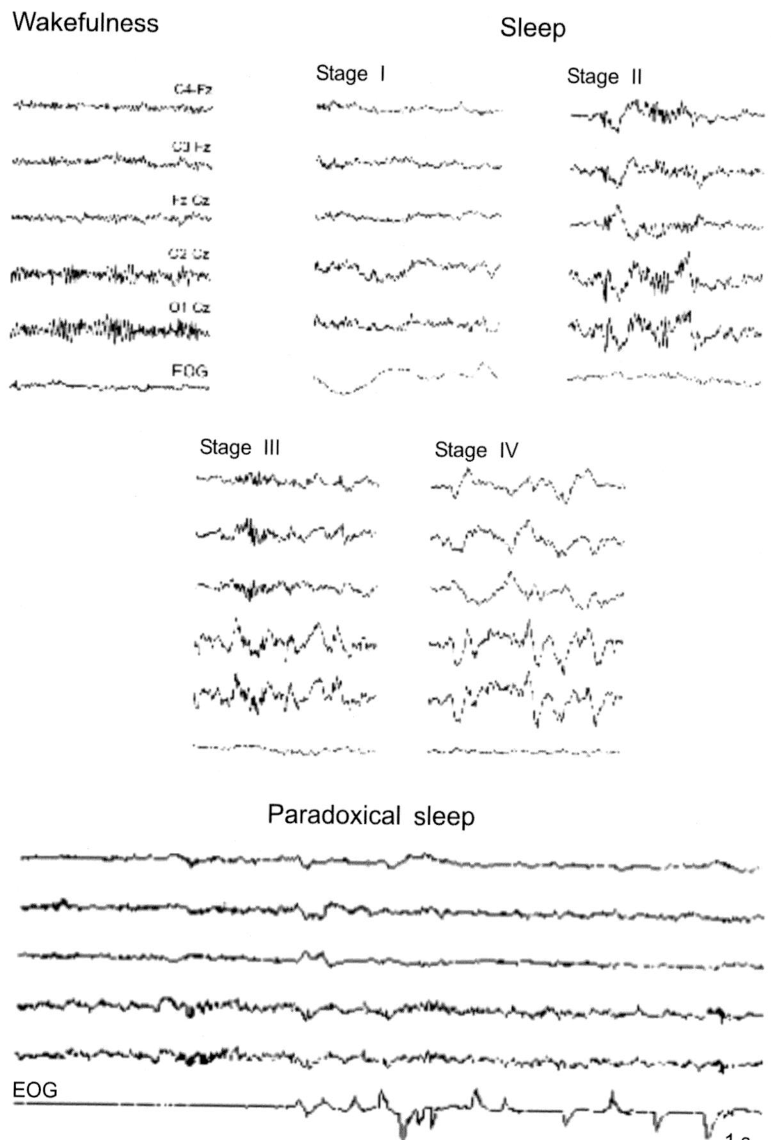

Figure 2.1 Brain bioelectrical activity during sleep and waking. The EEG and electro-oculogram (EOG) of a quiet episode of W are shown (note the α rhythm in the EEG). Sleep is divided into stages N1 to N3, by the characteristics waves, being the last the classical slow wave sleep (SWS). The paradoxical sleep (PS) represents about 25% of the total amount of sleep and is periodically associated with the presence of REM (EOG). Cal. 75 μV.

these variables, always associated with behaviour. The transition from one sleep stage to another is progressive and the polygraph elements change, with different temporal courses, until achieving completely the characteristics of the next stage. Wakefulness is a state in which a characteristic mental activity is developed (the vigil consciousness) accompanied by the execution of voluntary movements. It is not a homogeneous state but it is composed of numerous activity-rest ultradian cycles of about 90 minutes duration.

Since the discovery that electrical stimulation of a mesencephalic region provokes behavioural and electrophysiological awakening in animals, EEG activation, an ascending reticular activating system with a diffuse projection to the forebrain has been demonstrated (Moruzzi and Magoun, 1949; Moruzzi, 1963). However, this assumption was contradicted by several experimental approaches (Adametz, 1959; Chow and Randall, 1964; Yamamoto et al., 1988; John and Ranshoff, 1996). Two experimental lesions successive in time, separated by days, of the mesencephalic reticular formation do not produce a comatose state; rather, animals are able to feed normally and have sleep-waking cycles. Moreover, they can retain and produce conditioned responses after the second mesencephalic lesion, which indicates that there is no unique 'waking centre. Some rostral regions can hold the ability to maintain consciousness even after the mesencephalic lesions (John, 2001, 2006).

Genetic Aspects

Recent advances opened new perspectives in the genetic approach to normal sleep, although experimental results are scarce. The findings that certain disorders, such as fatal familial insomnia, are transmitted genetically and the role of hypocretins in narcolepsy (Chemelli et al., 1999) opened an avenue for the genetic approach to sleep pathologies. There are just a few genes whose mutations cause sleep disorders. Furthermore, disorders related to just one gene are rare; neither do we know the influence of environmental factors that may induce those genes' expression. The recently identified autosomal recessive gene that controls the frequency of the hippocampus theta rhythm during PS represents an example of genetic regulation of bioelectrical rhythms (Tafti, 2003; Dauvilliers et al., 2005).

Each component of sleep has to be considered as a complex phenotype. Studies in identical and fraternal twins have contributed to determination of possible environmental factors. A high similarity in sleep parameters among twins has been reported, thus suggesting an important contribution of genetic factors to the sleep organization (Dauvilliers et al., 2005). However, these studies are scarce and can still be deemed preliminary.

Studies have also identified hundreds of brain transcripts conserved across species that change their expression level between sleep and waking, signifying that the functional consequences of sleep are also shared among species. Thus specific genes can affect sleep, and conversely, sleep can influence gene expression in the brain (Cirelli, 2009).

Genetics of Sleep Pathology

- *Restless leg syndrome (RLS).* RLS is a common sleep disorder that is often characterized by periodic limb movements during sleep. RLS can be familial (autosomal dominant in up to a third of cases).
- *Obstructive sleep apnoea syndrome.* The genetic basis of this disorder, however, is difficult to study, because many of the risks factors, including obesity and alterations of the craniofacial morphology, are also under genetic control.
- *Narcolepsy.* Narcolepsy type 1 (NT1) is a chronic sleep disorder caused by a selective loss of hypocretin-producing neurons due to a mechanism of neural destruction that indicates an autoimmune pathogenesis (Martinez-Orozco et al., 2016). 85% of narcoleptic patients share an allele of chromosome. Another gene involved is the orexin gene already located (Guilleminault and Anagnos, 2005).

There is no doubt that more genes affecting sleep phenotypes will be identified in the future, and it is reasonable to assume that most mutations will have less striking effects on sleep in mammals than in simpler organisms, as mammalian genomes are redundant (Cirelli, 2009).

Sleep Deprivation

Human volunteers who underwent total sleep deprivation (over periods of up to 200 hours) presented signs of great fatigue, attention disorders and irritability, and a significant decrease in discrimination abilities. In some cases they presented hallucinations and balance, sight, and speech disorders. The wakefulness EEG in sleep-deprived subjects shows a decline in the alpha rhythm: Volunteers cannot keep this rest rhythm for more than 10 seconds. In addition, episodes of delta and theta waves become frequent. Selective deprivation of PS cannot be sustained for long, as microsleep episodes invade wakefulness uncontrollably, whereas total deprivation of 16 hours or longer in humans implies the loss of cognitive abilities, which was well demonstrated experimentally (Durmer and Dinges, 2005). Total sleep deprivation in rats revealed that these animals die within 15–22 days suffering from a general functional depression, presenting neurological and behavioural disorders.

Sleep structural parameters also found that relative to baseline sleep, volumetric changes were present after 12 or 24 h of training but in different directions, with increases in subcortical grey matter (GM) and white matter (WM) and decreases in the ventricles. All these volumetric changes were reversed by recovery sleep (Bernardi et al., 2016).

Physiological regulation in sleep

Physiological regulation in mammals depends on the ultradian wake-sleep cycle (Parmeggiani, 2005a,b). This dependency is the result of the changing functional dominance of phylogenetically different structures of the encephalon across the different behavioural states of the cycle. The functional similarity of physiological events during NREM sleep in different species and the variety and variability of such events during REM sleep within and between species define the characteristic differences between these states of sleep. Intrinsic nervous processes specific to the state of REM sleep may cause somatic and autonomic variability without relationship to mental content or homoeostatic control. The basic somatic features of NREM sleep are the assumption of a thermoregulatory posture and a decrease in antigravity muscle activity. The basic somatic features of REM sleep are muscle atonia, REM and myoclonic twitches. The basic autonomic feature of NREM sleep is the functional prevalence of parasympathetic influences associated with quiescence of sympathetic activity. The basic autonomic feature of REM sleep is the great variability in sympathetic activity associated with phasic changes in tonic parasympathetic discharge. In all species, the somatic and visceral phenomena of NREM sleep are indicative of closed-loop operations automatically maintaining homoeostasis at a lower level of energy expenditure compared with quiet W. In contrast, the somatic and visceral phenomena of REM sleep are characterized in all species by the greatest variability: This is a result of open-loop operations of central origin impairing the homoeostasis of physiological functions (poikilostasis). The demonstration, in terms of reactive homoeostasis, of different functional states of the ultradian sleep cycle that are characterized by either homoeostasis (NREM sleep) or poikilostasis (REM sleep) of physiological functions is based on the criterion of short-latency stimulus-response relationships. This basic functional dichotomy applies to the nervous control of body

(*Continued*)

(Continued)
temperature and of circulatory and respiratory functions. In contrast, gastrointestinal, endocrine and renal functions do not fit this criterion. For example, many aspects of gastrointestinal function are not constrained within the temporal boundaries of single sleep states and appear, at most, to be modulated by changes in the autonomic nervous system outflow during sleep. On the other hand, there are changes in endocrine secretion that are specific to a single sleep state. However, such changes are the result of ultradian or circadian modulation rather than of a homoeostatic response to exogenous or endogenous disturbances in terms of reactive homoeostasis.

Pier Luigi Parmeggiani
Universita di Bologna, Bologna, Italy (2008)

SLEEP AND ITS ASSOCIATED PHYSIOLOGICAL CHANGES

During the last decades, it has been demonstrated that all physiological functions vary depending on the moment of sleep-wakefulness cycle, and many of these functions are associated with a particular sleep stage. Thus the autonomic control expressed through cardiovascular and respiratory manifestations is modified concomitantly with the SWS-PS sequence. The sensory processing, the endocrine function, and others analyzed later in this chapter are also interrelated with the sleep-wakefulness cycle. Furthermore, during PS the homoeostasis is transitorily neglected (Parmeggiani, 1980).

Sensory Information Processing

Information from the outer world and the body (inner world), both conscious and unconscious, enters through sensory receptors and is continuously evaluated in the CNS. This information connects the individual with the environment and also keeps the brain in close contact with the sleeper's internal medium, viscera, muscles, joints and so on (Velluti, 1997). I want to stress now that the information about the 'inner world' is continuously processed in sleep and waking by the constantly working CNS (Velluti et al., 2000; Velluti and Pedemonte, 2002, 2012).

During sleep, psychomotor reactions to environmental stimuli are clearly reduced. We find ourselves relatively isolated from the environment. However, from an electrophysiological point of view, the evoked potentials or the unitary responses of certain neuronal groups are comparatively greater during SWS than during wakefulness. This fact, apparently paradoxical, was demonstrated for visual and auditory information. The thalamic and cortical auditory evoked potentials exhibit greater amplitude during SWS when compared to wakefulness and PS (Campbell et al., 1992; Coenen, 1995; Velluti, 1997; Bastuji and García-Larrea, 1999, 2005).

The auditory system is a telereceptor system that remains relatively 'open' during sleep. The possibility of establishing auditory contact with the external world must have been important, from a phylogenetic point of view, for the survival of the most vulnerable species, allowing them to wake up and develop an adequate reaction before a predator strikes. At present, we keep using the auditory system to provoke our daily awakening. Much experimental data nowadays support the idea that the entire auditory system, from the cochlear receptor to the cortical neurons, keeps processing information during sleep, although differently from during wakefulness. Shifts in the neuronal discharges correlated with sleep stages have been verified along the whole pathway and would possibly be the electrophysiological expression of the changes in sensory processing that take place while we sleep. Thus we are able to perceive an auditory stimulus, process it and compare it to information stored in our memory and make decisions such as waking up or continue sleeping.

These results suggest that neurons belong to different neuronal networks that participate in multiple processing besides the specific sensory one, being the general state of the brain being the one in charge of modulating this activity (Pedemonte and Velluti, 2005a,b; Velluti, 2005; Velluti and Pedemonte, 2012). Inversely, the complete absence of auditory input produces modifications in sleep and wakefulness, increasing both sleep phases to the detriment of wakefulness (Pedemonte et al., 1996).

The sleeping brain imposes conditions for the arrival and processing of auditory sensory information; furthermore, it is my tenet that what is observed in the auditory system can be valid for all the sensory systems. Therefore sensory activity arriving in the CNS sleep during early development, which in human neonates mostly takes place during sleep, is a relevant factor in the 'sculpting' or maturing of the brain. The sensory information processed during sleep in an early stage of life (days, months) must participate in the CNS maturation, since sensory information keeps entering while we sleep.

Sleep and Dreams

Current physiological studies demonstrate that dreams are regularly present in all phases of sleep, stages II, SWS III–IV and PS, reflecting a more or less complex series of sensory and emotional events (Portas, 2005). However, dreams that take place during the last cycles of PS, just prior to awakening, would be the easiest to recall and thereby the most frequently reported. It is possible to obtain reports of dreams in 85%–90% of the awakenings provoked during PS and in 50% of those provoked during stage II and SWS stages III–IV (Foulkes, 1962; Cicogna et al., 1991).

The idea that PS dreams are bizarre, unreal, hallucinating and so on seems to result from the scarce amount of studies on which this tenet was based. Properly controlled analyses performed in the sleep laboratory show that during PS there are also dreams associated with common experiences of daily life. Recent research states that the content of those dreams recorded in SWS and PS are equal as long as they have equivalent duration. These concepts lead to postulate the existence of one single dream generator system, which functions throughout the diverse sleep phases. In addition, it is important to remark the coherence and thematic organization of every dream, which reveals that dreams are the product of a brain working in an organized fashion. Some authors have a divergent viewpoint to the proposed dichotomy between PS and stage II and SWS dreams, sleep mentation (Hobson et al., 1998). Others researchers consider the differences as quantitative instead of qualitative (Cicogna et al., 1991).

Concerning the thematic content of dreams, 100% of the narrations report visual images and 65% report auditory sensations (McCarley and Hoffman, 1981), whereas the percentages associated with other sensory modalities are significantly lower. Depending on the sensory system involved, information from the environment can 'enter' a dream to become part of the report. The importance of the auditory system, which remains constantly 'open', lies in the possibility to continuously control environmental sounds (Fig. 2.2). Although less frequent than sensations, emotions are also expressed in dreams, anxiety being the most often reported. Dreams can repeatedly produce changes in heart and respiratory rates as a result of the autonomic system activation.

Motor activity is also present during dreams although significantly reduced by the motor neuron inhibition present during sleep (Pompeiano, 1967; Lai and Siegel, 1991). The spinal motor neuron inhibition in PS is opposed to the increased bursting firing of pyramidal tract fibres (Evarts, 1964).

Although the specific sources of dreams remain an enigma, we can generally state that they are constitute (1) what the individuals have in

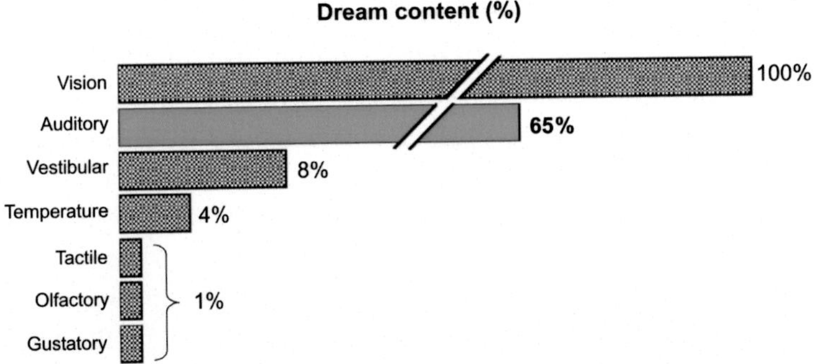

Figure 2.2 Sensory modalities present in the dreams content. The dream visual images (100%) together with auditory 'images' (65%) are the most salient components of dreams. *Modified from McCarley, R.W., Hoffman, E.A., 1981. REM sleep dreams and the activation-synthesis hypothesis. Am. J. Psychiatry 38, 904–912.*

their memory; (2) sensory information from the outer and inner world, -unconscious intrusions while the dream is taking place; (3) possible genetically transmitted information and (4) prior information that occurred during the previous W.

The fact that depressed patients have dreams with depressive characteristics and schizophrenic patients develop disorganized dreams point to a continuation between the psychical activity of wakefulness and the oneiric activity. But there is at least one objective difference: Dreams focus on only one oneiric experience, while during wakefulness it is possible to sustain multiple elements consciously and simultaneously.

Dreams were also interpreted as a rehearsal for brain activity involved in relevant behaviour (Cartwright, 1974). However, it was also postulated that dreaming may be an accidental byproduct of brain activity; besides, Jouvet (1999) proposed that, during dreaming, the brain may be genetically 'reprogrammed'.

Dreaming and sleep mentation reflect a more or less complex sequence of sensory and emotional events produced by the sleeper's mind, associated to some episodes of SWS (NREM) sleep, being present in 90% of PS (REM) sleep episodes.

Monitoring brain activity using high-density EEG, participants were awakened during REM and non-REM sleep and reported on the presence or absence of dreaming. The authors found that the presence of high-frequency EEG activity was predictive of dreaming during non-REM sleep (Siclari et al., 2017).

Cardiovascular Functions

During sleep, the mean arterial pressure decreases as a result of a drop in the diastolic and systolic blood pressures. The lowest value is recorded during stages III–IV of SWS. In humans, the PS blood pressure becomes variable and exhibits transient increases of up to 40 mm Hg, which overlap with a tonic hypotension. These blood pressure increments coincide with PS phasic events (REMs, muscle twitches). The pulmonary artery blood pressure remains stable during all sleep phases (see Silvani and Lenzi, 2005).

The heart rate decreases during SWS, predominantly in stages III–IV. During PS it becomes variable mainly during the phasic activity. The cardiac output becomes moderately reduced during both SWS and PS, constituting another element that contributes to the previously mentioned blood pressure drop. Sleep hypotension also depends on vasodilatation. During PS, there are periods of vasoconstriction in the skeletal muscles, which may be the cause of the phasic increments in blood pressure.

Cerebral Blood Flow

Studies in human subjects using diverse techniques have revealed regional raises or drops in blood flow during SWS. Studies in humans and other species agree in the existence of a significant rise in cerebral blood flow (CBF) during PS, with phasic increases overlapping the tonic increase. The mechanisms responsible for these changes have not been clearly characterized, although existing data suggest that they might be induced by local metabolic variations (Reivich, 1974; Braun et al., 1997).

The recording of cerebral oxygen in cats exhibits a particular distribution of O_2 availability during PS, which has been denominated the *pO_2 PS system* (Velluti, 1985). This system includes the reticular formation, the *reticularis pontis oralis*, the basal forebrain, the hypothalamus, the amygdala and the cerebellum. Oxygen availability in the cortex does not show changes during PS, presumably because its blood supply has a better distribution than subcortical structures or because of the cortical laminar arrangement could prevent the recording of the pO_2 oscillating pattern (Fig. 2.3).

This oscillating pattern, which was attributed to a decrease in the local homoeostatic control, also has been interpreted as an increase in glucose degradation induced by anaerobic metabolism during neural activity

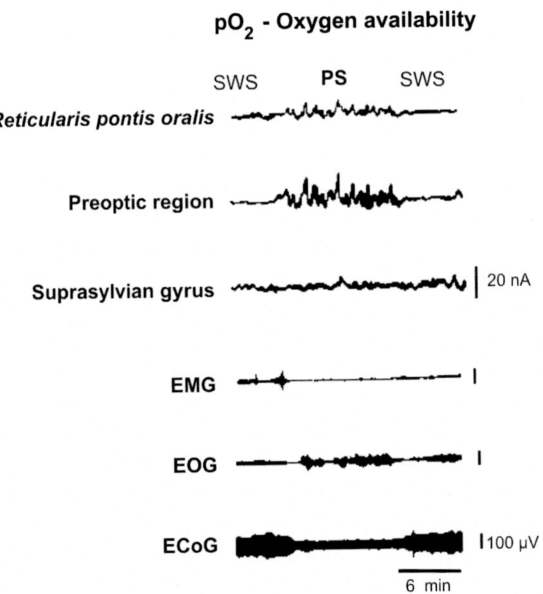

Figure 2.3 Oxygen cathodes implanted in the *reticularis pontis oralis*, the preoptic region and the suprasylvian gyrus recorded local oxygen availability during SWS and PS, in cats. The upper two traces show high-amplitude oscillations, which were not observed in the cortex. ECoG, electrocorticogram.

increments (Velluti et al., 1965; García-Austt et al., 1968; Velluti and Monti, 1976; Velluti et al., 1977; Velluti, 1985, 1988; Franzini, 1992).

Neurogenic vasomotricity has also been involved in the CBF regulation, although this aspect has not been sufficiently studied so far. The most widely accepted theory nowadays is the coupling of blood flow and neuronal activity by metabolic factors. Experimental data over the last decade, which show a global reduction in the CBF during the night with flow values postsleep significantly lower than flow values presleep, would support the classical hypothesis of the 'restorative' function of sleep (Zoccoli et al., 2005). Once sleep requirements have been fulfilled, resuming the operational brain level would take place at a lower metabolic cost during wakefulness subsequent to sleep. Throughout wakefulness, a 'debt of sleep' would gradually be created, manifested as a growing propensity to sleep, which would be 'repaid' during sleep. Since SWS stages are longer at the beginning of the night and decline exponentially in the course of the hours with a temporal course similar to that of the CBF, some authors speculate about a functional association between both; that is, the CBF diminishes over the night and slow waves gradually decline.

Imaging

Studies performed in humans using positron emission tomography (PET) have opened a new avenue for the research of sleep processes. Maquet (2000) showed, through the use of PET with *2-deoxy-D-glucose*, a 12% decrease of the cerebral glucose metabolism during SWS compared to wakefulness, while PS produced a general increase of 16% (Fig. 2.4).

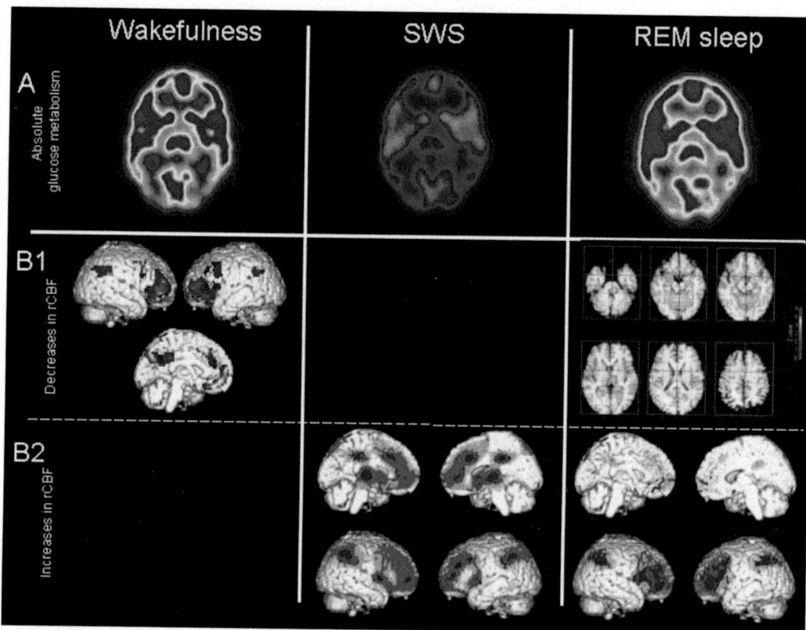

Figure 2.4 Glucose metabolism and regional cerebral blood flow (CBF) during wakefulness (W), slow wave sleep (SWS) and paradoxical sleep (PS, REM). (A) Cerebral glucose metabolism quantified in the same individual at 1-week intervals, using fluorodeoxyglucose and PET. There is a significant decrease in the average glucose metabolism during SWS compared to W. During PS (REM) the glucose metabolism is as high as during W (Maquet et al., 1990). (B1) Distribution of the highest brain activity, assessed by CBF measurement using PET during W and PS (REM sleep). The most active regions during W are located in the associative cortices in the prefrontal and parietal lobes (Maquet, 2000). During PS (REM) the most active areas are located in the pontine tegmentum, the thalamus, the amygdaloid complexes and the anterior cingulate cortex (Maquet et al., 1997). (B2) Distribution of the lowest regional brain activity during SWS and PS (REM sleep) using the same method as in B1. In both sleep stages, the least active regions during W are located in the associative cortices in the prefrontal and parietal lobes. During SWS, the brainstem and thalamus are particularly deactivated. *Modified from Maquet, P.A.A., Sterpenich, V., Albouy, G., Dang-vu, T., Desseilles, M., Boly, M., et al., 2005. Brain imaging on passing to sleep. In: Parmeggiani, P.L., Velluti, R.A., (Eds.), The Physiologic Nature of Sleep. Imperial College Press, London, pp. 489–508.*

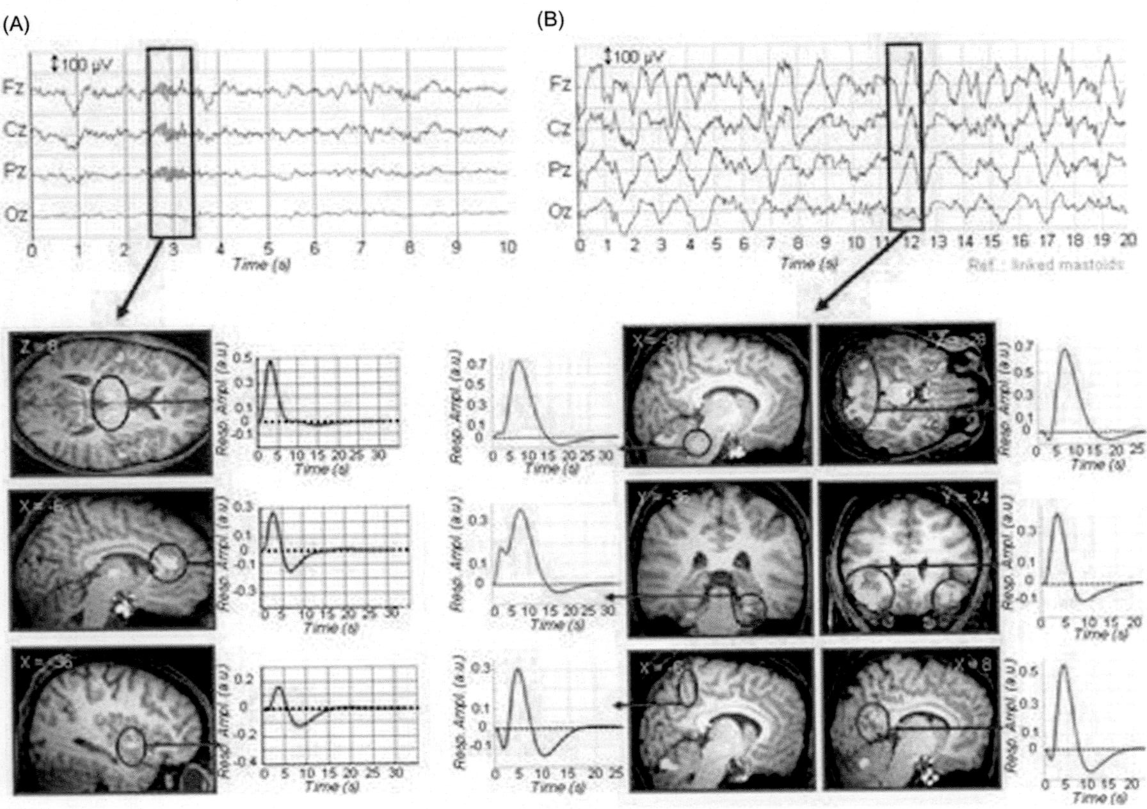
(A)
(B)
100 μV
Fz
Cz
Pz
Oz
Time (s)
Ref.: linked mastoids
Z = 8
Y = 24
X = 8
Resp. Ampl. (a.u)
Time (s)

Fig. 2.5 shows neural correlations among EEG spindles and slow waves of NREM sleep oscillations and fMRI (Dang-Vu et al., 2010).

Experimental evidence demonstrates that cortical activity, in addition to participating in the global changes that characterize the electrophysiological configuration of sleep, is also involved in specific processes; it is reactivated during the SWS and PS episodes subsequent to a new training (Louie and Wilson, 2001; Lee and Wilson, 2002).

The regional CBF measurements using PET carried out on human subjects has permitted demonstrating that the least activated regions during SWS are localized in the dorsal pons, mesencephalon, cerebellum, thalamus, basal ganglia, hypothalamus and the prefrontal cortex, which indicates that the rest of the brain remains significantly active during this stage. During PS, the especially active areas are the pontine *tegmentum*, some thalamic nuclei, the amygdaline complex, the hippocampus, the cingulate cortex and the posterior temporo-occipital cortices (Maquet et al., 2005). Such promising studies, are still in progress and should not lead us to fractionate the brain, as happened in the late 19th century during the height of phrenology.

◀ **Figure 2.5** Neural correlates of NREM sleep oscillations as evidenced by fMRI. (A) fMRI correlates of spindles. The upper panel shows a (stage N2) NREM sleep epoch depicting a typical spindle on a scalp EEG recording. Brain activity is estimated for each detected spindle compared to the baseline brain activity of NREM sleep. The lower left panels shows the significant brain responses associated with spindles ($P < 0.05$, corrected for multiple comparisons on a volume of interest), including the thalamus, anterior cingulate cortex and insula (from top to bottom). Functional results are displayed on an individual structural image (display at $P < 0.001$, uncorrected), at different levels of the x, y, z axes as indicated for each section. The lower right panels show the time course (in seconds) of fitted response amplitudes (in arbitrary units) during spindles in the corresponding circled brain area. All responses consist in regional increases of brain activity. (B) fMRI correlates of slow waves. The upper panel shows a (stage N3) NREM sleep epoch depicting typical slow waves on a scalp EEG recording. Brain activity is estimated for each detected slow wave compared to the baseline brain activity of NREM sleep. The lower panels shows the significant brain responses associated with slow waves ($P < 0.05$, corrected for multiple comparisons on a volume of interest), including the brainstem, cerebellum, parahippocampal gyrus, inferior frontal gyrus and precuneus and posterior cingulate gyrus (from left to right and top to bottom). Functional results are displayed on an individual structural image (display at $P < 0.001$, uncorrected), at different levels of the x, y, z axes as indicated for each section. The lower side panels show the time course (in seconds) of fitted response amplitudes (in arbitrary units) during slow wave in the corresponding circled brain area. All responses consist of regional increases of brain activity (Dang-Vu et al., 2010).

Finally, the activity of the entire CNS is necessary to generate and maintain sleep; therefore, both the regions that become activated and those reducing their activity are in effect necessary to achieve a sleeping brain.

Recovery sleep was associated with a significant cortical thickness increase. Changes in cortical thickness have been observed mainly during development, aging, or in response to chronic manipulations (Bernardi et al., 2016).

Respiratory Changes

During wakefulness, respiration is controlled by a triple mechanism: (1) the metabolic, which ensures the homoeostasis of arterial O_2 and CO_2 through information conveyed from the central and peripheral chemoreceptors; (2) voluntary control, which permits coordinating ventilation and other functions such as speech and cough; (3) the action of spinal motor neurons, which innervate the respiratory muscles and receive tonic inputs that help to maintain their membrane potential level at a certain degree of depolarization. In addition, the spinal motor neurons receive innervation from respiratory centres of the brainstem, which have a semiautomatic function (Fig. 2.6). The metabolic control is integrated at the bulboprotuberantial level and finally provokes the necessary changes to activate the respiratory muscles. The voluntary control most certainly involves structures that control brainstem and spinal respiratory neurons, such as the amygdala, the periaqueductal gray and frontal cortex (Orem et al., 2002; Orem, 2005).

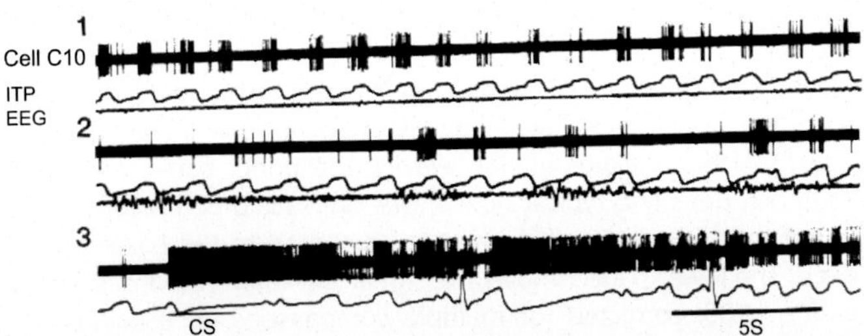

Figure 2.6 Medulla inspiratory cell sensitive to sleep. (1) Spontaneous activity of the cell during wakefulness (W). Top trace, action potentials of the cell; middle trace, intratracheal pressure (ITP, negative pressures indicated by upward deflections); lower trace, EEG. (2) Spontaneous activity of the cell during drowsiness or slow wave sleep (SWS). The activity of the cell decreased in drowsiness and SWS. (3) Intense activation of the cell during and after behavioural inhibition of inspiration elicited by a conditioning stimulus (CS). *Modified from Orem (1989).*

The respiratory changes that take place during SWS and PS reflect the predominance of the metabolic control during SWS and a decrease in this control during PS. A characteristic nonhomeostatic change is present in PS, when ventilation no longer depends on the metabolic control.

Both drowsiness and stage II provoke an unstable respiratory rhythm with consecutive hypoventilations and hyperventilations, called *periodic ventilation*. During stages III–IV ventilation becomes regular, with a higher amplitude and a lower respiratory rate and a mild decrease in the per-minute volume. This is associated with a decrease in the metabolic rate and also with variations in the central control of respiration. As SWS starts, automatic control mechanisms are released, with the inactivation of the telencephalic mechanisms that command breathing during wakefulness. During stages I and II, episodes of respiratory instability alternate with periods of regular breathing, which will later stabilize in a regular breathing rhythm by stages III–IV. The partial pressure of alveolar CO_2 increases, whereas the partial pressure of alveolar and arterial O_2 decreases. The chemoreceptors' response to CO_2 is moderately reduced while the response to hypoxia does not change. Both respiratory rate and depth remain relatively constant.

The respiratory rhythm during PS is typically faster and irregular, exhibiting episodes of apnoea and hypoventilation. The muscle hypotonia contributes by decreasing the strength of the chest expansion and increasing the resistance to airflow in the upper air pathways. The diaphragm maintains an irregular activity and atonia of intercostal muscles occurs in PS in cats (Parmeggiani and Sabattini, 1972). While Pompeiano (1967) proposed that atonia is the result of active inhibition, both excitatory and inhibitory tonic processes affect the respiratory system in PS (see the review by Orem, 2005).

Endocrine Functions

Blood levels of every hormone exhibit cycles. These cycles are modulated by the circadian rhythm of light and darkness, sleep and wakefulness, autonomic activity, and the like, all functions that converge on the hypothalamus. The hypothalamus-hypophysis axis is the link between sleep processes and hormonal secretion.

Many hormones are secreted with a rhythm set by the wakefulness–sleep cycle. The influence of the sleep-wakefulness cycle on the endocrine system has been subdivided into three basic types: (1) hormones modulated by a

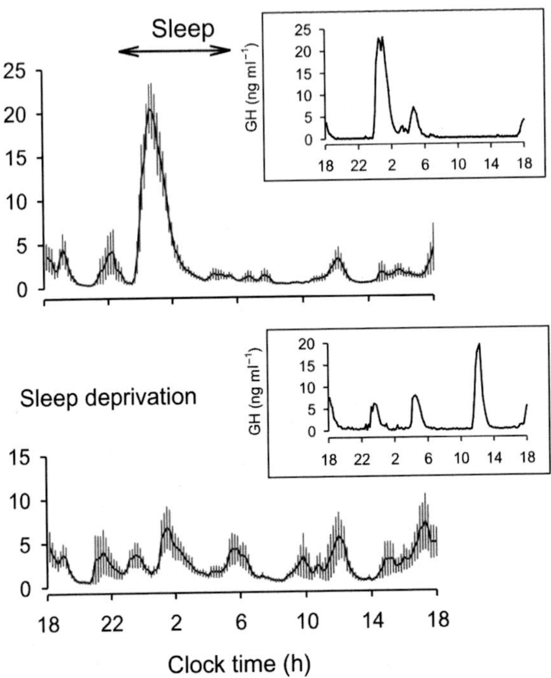

Figure 2.7 Effect of sleep deprivation on mean 24-hour growth hormone (GH) levels in 10 subjects. In sleep-deprived persons, the reduction of the sleep-related pulse is compensated for by the emergence of large individual pulses during the day, so that the amount of GH secreted over 24 hours is similar whether or not a person had slept during the night. Insets give a typical example of individual profiles. Bars indicate standard error of the mean. *Modified from Brandenberger, G., 2005. Endocrine correlates of sleep in humans. In: Parmeggiani, P.L., Velluti, R.A. (Eds.), The Physiologic Nature of Sleep. Imperial College Press, London, pp. 433–453.*

particular stage of sleep, such as the peak in the growth hormone (GH; Fig. 2.7) levels during a specific phase of SWS; (2) hormones highly influenced by the whole period of sleep, such as prolactin and thyrotrophin (TSH); and (3) hormones weakly modulated by sleep, such as ACTH (adrenocorticotropic hormone), cortisol and melatonin, despite exhibiting strong circadian rhythms (Brandenberger, 1993, 2005).

The GH, essential for development, shows a secretion cycle closely linked to the first episode of sleep stage IV (Takahashi et al., 1968; Sassin et al., 1969). If SWS is delayed or prevented, this hormone will not be secreted in appreciable quantities, whereas, if the SWS is facilitated, by physical exercise for example, this hormone will be released in high quantities. In infants and adults with prolonged SWS, secretory episodes of great magnitude can be obtained. In elderly people, coincidentally with the normal decrease of SWS, this hormone's secretion is greatly reduced. Episodes of PS take place mainly during the descending phase of the secretory pulses of GH or in its nadir.

Melatonin exhibits a secretory peak at the beginning of the night, generated by a central pacemaker located in the suprachiasmatic nucleus of the hypothalamus, which receives projections from the retina. Its major function is the coordination of circadian rhythms. It is considered a chemical code through which the brain 'understands' that it is nighttime; the longer the night, the longer is the melatonin secretion, constituting in some species the temporal key to cycle seasonal rhythms (Cardinali, 2005).

Finally, all findings seems to indicate that the endocrine system pulsatility is commanded by complex ultradian clocks. It is important to mention that the functional significance of the diverse temporal relations between sleep and hormones is still unknown.

Physiological Regulation in Sleep

Physiological regulation in mammals depends on the ultradian wake-sleep cycle. This dependency is the result of the changing functional dominance of phylogenetically different structures of the encephalon across the different behavioural states of the cycle. The functional similarity of Physiological events during nonrapid-eye-movement SWS (NREM) sleep in different species and the variety and variability of such events during rapid-eye-movement PS (REM) sleep within and between species define the characteristic differences between these states of sleep. Intrinsic nervous processes specific to the state of REM sleep may cause somatic and autonomic variability without relationship to mental content or homoeostatic control. The basic somatic features of SWS (NREM) sleep are the assumption of a thermoregulatory posture and a decrease in antigravity muscle activity. The basic somatic features of PS (REM) sleep are muscle atonia, REMs and myoclonic twitches. The basic autonomic feature of NREM sleep is the functional prevalence of parasympathetic influences associated with quiescence of sympathetic activity. The basic autonomic feature of REM sleep is the great variability in sympathetic activity associated with phasic changes in tonic parasympathetic discharge. In all species, the somatic and visceral phenomena of NREM sleep are indicative of closed-loop operations automatically maintaining homoeostasis at a lower level of energy expenditure compared with quiet wakefulness. In contrast, the somatic and visceral phenomena of REM sleep are characterized in all species by the greatest variability: This is a result of open-loop operations of central origin impairing the homoeostasis of Physiological functions (poikilostasis). The demonstration, in terms of reactive homoeostasis, of different functional states of the ultradian sleep cycle that are characterized by either homoeostasis (NREM sleep) or poikilostasis (REM sleep) of Physiological functions is based on the criterion of short-latency stimulus-response relationships. This basic functional dichotomy applies to the nervous control of body temperature and circulatory and respiratory functions. In contrast, gastrointestinal, endocrine and renal functions do not fit this criterion. For example, many aspects of gastrointestinal function are not constrained within the temporal boundaries of single sleep states and appear, at most, to be modulated by changes in the autonomic nervous system outflow during sleep. On the other hand, there are changes in endocrine secretion that are specific to a single sleep state. However, such changes are the result of ultradian or circadian modulation rather than of a homoeostatic response to exogenous or endogenous disturbances in terms of reactive homoeostasis.

Body Temperature

In homoeothermic animals, the interaction between hypothalamic and cortical mechanisms controls body temperature. In SWS the automatic mechanisms are released from the cortical control. During PS temperature regulation is interrupted. The decrease in the muscular tone and the shivering absence reduces the body's ability to produce heat. Body temperature falls along the night to reach the lowest levels by the last hours of sleep. There is a rise in skin temperature coincident with PS periods. Sleep continuity is altered by ambient conditions (Fig. 2.8; Libert and Bach, 2005).

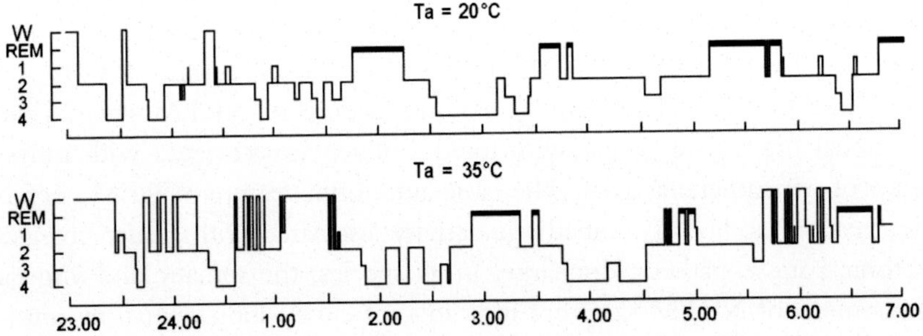

Figure 2.8 Altered hypnogram of an adult person sleeping at an ambient temperature of 20°C and 35°C. *Modified from Libert and Bach (2005).*

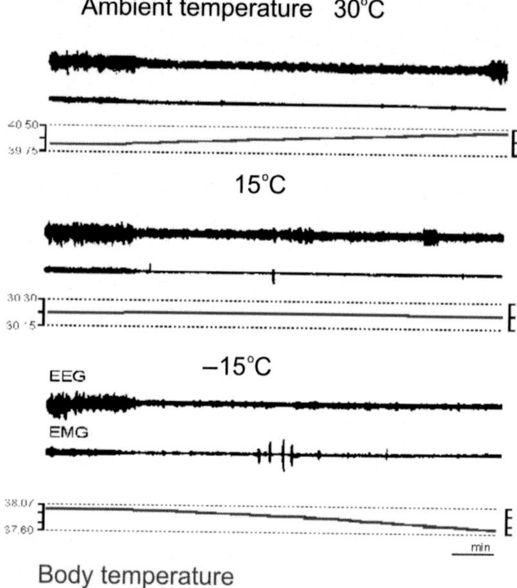

Figure 2.9 Body temperature shifts during PS in cats. Exposed to a high ambient temperature (30°C) the body increased its temperature. A moderate ambient temperature (15°C) is associated with stable body temperature. When the ambient temperature is very low (−15°C), the body also decreased its temperature. *Modified from Parmeggiani, P.L., 1980. Temperature regulation during sleep: a study in homeostasis. In: Orem, H., Barnes, Ch. (Eds.), Physiology in Sleep. Academic Press, New York, pp. 98–136.*

Works carried out by Parmeggiani (1980, 2005a,b) during the last decades have demonstrated that the correlation between body temperature and room temperature is variable and it depends on the moment of the wakefulness–sleep cycle being analyzed. A cat exposed to changing room temperatures is able to control and maintain its body temperature constant during wakefulness and SWS, whereas during PS its body temperature will rise or drop if its environment is warm or cold, respectively. Hence, we can assume that during PS the animal enters a state similar to that of a poikilotherm (Fig. 2.9).

Changes in Other Functions

- *Renal function.* Glomerular filtration, urine volume and sodium, potassium and calcium excretion decrease during sleep. Urine concentration is higher during sleep than wakefulness because of the increment of the antidiuretic hormone.

- *Digestive functions.* Studies have shown a decrease in gastric acid secretion during sleep in normal humans. Subjects suffering from duodenal ulcers have the acid secretion permanently increased over the sleep and wakefulness cycle. Recordings of bowel motility show conflictive changes up to the present time, although the oesophagus motility is consistently reduced.
- *Sexual functions.* Penile erection or tumescence have been shown in human subjects between 3 and 79 years of age during PS; although present in adolescents, this is not only confined to this stage. Although its functional role remains unknown, the presence or absence of erection during sleep is employed for the differential diagnosis between organic and psychogenic impotence. Women exhibit clitoral erections and an increment of the vaginal blood flow during PS.

Homoeostasis

The existence of homoeostatic mechanisms that regulate all functions was already postulated by Cannon in 1929. However, it has recently been proposed that these mechanisms may be overcharged beyond their limits during certain states, such as extremely active wakefulness. Such deviations from the homoeostatic range can normally be compensated during wakefulness, restoring the functional equilibrium.

The study of sleep has demonstrated the existence of a functional disruption of the mechanisms of control, which are present during SWS and absent during PS. As a result, many of the basic functions, such as the blood pressure, breathing and temperature, are left out of the strict homoeostatic control and their values shift. As explained previously, the body temperature becomes dependent on the room temperature and the animal turns into a transitory poikilotherm (Parmeggiani, 1980). Furthermore, in certain regions of the cat's brain, the oxygen local availability becomes unstable during PS, which may also reflect an escape from homoeostatic regulation (Velluti, 1985; 1988; Velluti and Monti, 1976; Velluti et al., 1977). We cannot yet explain the reason for the existence of a sleep stage that lacks homoeostatic control. Which are the functions of that require such conditions?

The issue of an impairment of homoeostatic regulation during sleep was first raised by study of thermoregulation in cats exposed to negative (cold) and positive (warm) ambient thermal loads with respect to the neutral ambient temperature of the species. The observation (Fig. 2.10)

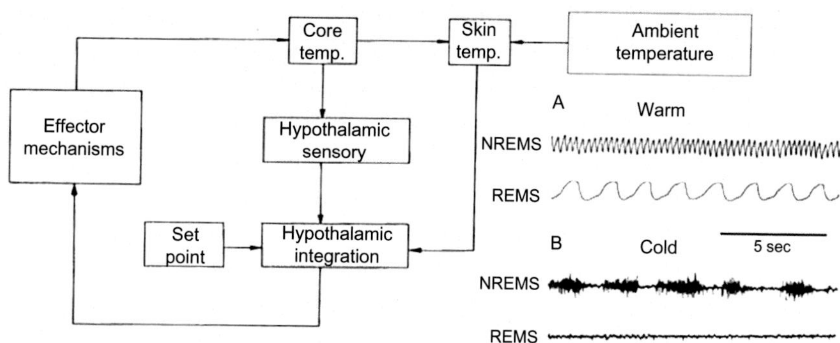

Figure 2.10 Tachypnea and shivering in cat sleep. (A) At high ambient temperature (37°C), the pneumogram shows tachypnea during slow wave sleep (SWS, NREM) and (B) the tachypnea absence during paradoxical sleep (PS, REM). At cold ambient temperature (6°C.) shivering is present only during SWS (NREM) and totally absent during PS (REM). *Modified from Parmeggiani, P.L., 2011. Systemic homeostasis and poikilostasis. Is REM Sleep a Physiological Paradox? Imperial College Press, London.*

was that in the cat two thermoregulatory responses were observed. Tachypnea and shivering are present in SWS sleep and absent in PS sleep (Parmeggiani, 2011).

The Energetic Metabolism

It is outstanding that evolution has permitted the appearance of the energy costly and dangerously unregulated PS. Whereas the lactate produced by the glycolysis in astrocytes would be capable of ensuring the ATP production necessary for the preservation of the active wakefulness and the PS (not clear; neurons also produce ATP), the SWS would be a stage for saving energy. However, it is difficult to study the energetic metabolism during the wakefulness–sleep cycle due to the absence of specific pharmacological tools that permit the performance of in vivo studies (Franzini, 1992; Cespuglio et al., 2005).

NEUROPHYSIOLOGY

Sleep, formerly considered a passive phenomenon, is nowadays recognized as actively generated by the brain. Experimental results such as the anatomy of lesions produced by a probable craniopharyngioma compressing the anterior hypothalamus, observed in Montevideo and published in Paris by Soca (1900), the lesions of the lethargic encephalitis published by von

Economo (1930), and the demonstration of a hypnogenic thalamic region (Hess, 1944) represent pioneer studies that lend support to this concept.

Consistent with this concept is the demonstration that during sleep the neuronal discharge in a number of regions increases. Moreover, neuronal activity changes, increasing or decreasing, along the different stages of sleep (McGinty and Szymusiak, 2003). Furthermore, a neuron that decrease its firing entering into sleep should be considered as also participating in the ongoing sleep processes. Unitary recordings of motor neuron fibres and other regions show that, contrary to what occurs during general anaesthesia, the discharge rate of some neurons increases or changes their firing pattern during sleep, reaching even higher levels than those of quiet wakefulness (Evarts et al., 1962). That sleep increases firing neurons was particularly observed in the preoptic region of the anterior hypothalamus (McGinty and Szymusiak, 2003). Fig. 2.11 shows two units, the auditory cortex and the preoptic one, belong to the same sleep-related network, perhaps organized by a common hub neuron(s).

The neuronal firing in the auditory system can increase and also decrease firing during sleep, perhaps participating in some process of sleep general organization (Fig. 10.2). In Chapter 5, Auditory Unit Activity in Sleep, the auditory neurons' sleep-related behaviour is shown.

Sleep, a cyclic and reversible stage, appears then as a unique physiological state of vigil consciousness abolition and reduction of response to the environment, which is accompanied by changes in multiple functions including oneiric 'consciousness'.

General Neural Approaches

Once the concept of sleep as an active process was established, the search for one or more sleep generating centres began (Kleitman, 1963). Abundant experimental evidence emerged from brain lesions, which led to the finding of regions involved in the generation or maintenance of sleep (Jouvet, 1962).

However, the sleep is not a function but it is instead another physiological state of the CNS and the body, which turns chimerical the search for a sleep single centre.

Bremer (1935) performed the first experimental preparation called *encephale isolè*, by sectioning of the CNS at the level of the first cervical segment, demonstrating that animals showed EEG and pupilar signs of

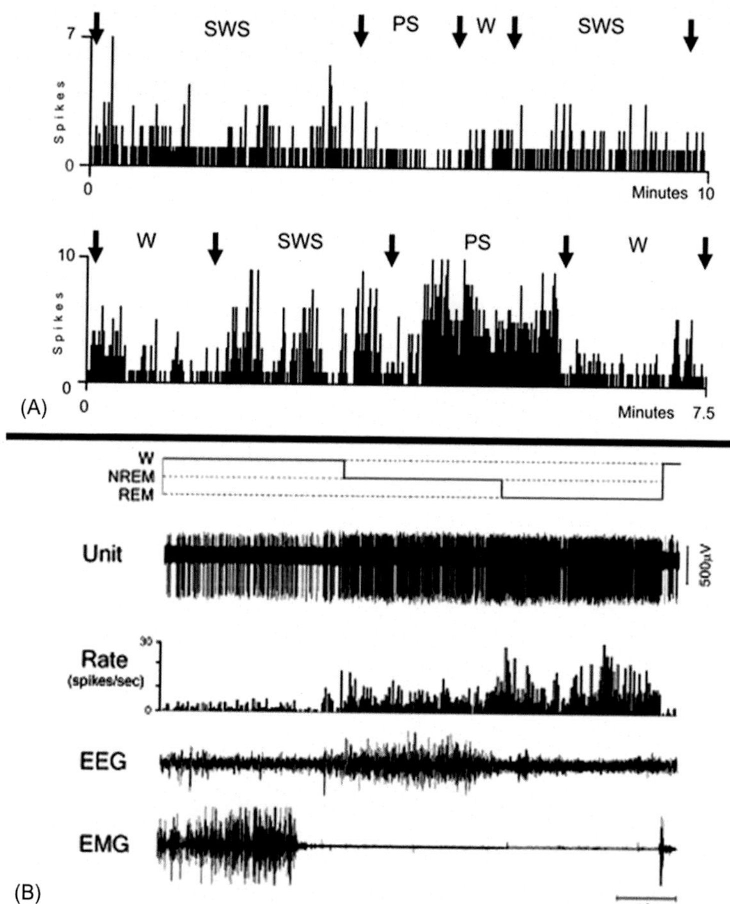

Figure 2.11 Comparison of two different units discharges during the sleep-waking cycle. A, two auditory cortical guinea pig neurons (A1) activity. Upper plot, a spontaneous discharge as a function of time. After fluctuating during slow wave sleep (SWS), the firing rate markedly decreases during paradoxical sleep (PS). The number of spikes is quantified over 50 ms epochs, 10 minutes of continuous recording. Lower plot, discharge as a function of time. The firing rate shows peaks during SWS and a quasitonic increase, on passing to PS. The number of spikes is quantified over 450 ms epochs, 7.5 minutes of continuous recording of wakefulness (W), SWS and PS, (Modified from Peña et al., 1999). B, discharge of a 'sleep-ON' or 'sleep-related' neuron during W, SWS and PS, recorded in the median preoptic nucleus of an unrestrained rat. Its firing rate is low during waking, increases at sleep onset and during SWS, and reaches even higher levels in PS (Modified from Suntsova et al., 2002). Both units, the auditory cortex and the preoptic one, belong to the same sleep-related network, perhaps organized by a common hub neuron(s).

wakefulness and sleep. In contrast, sectioning at the level of the mesencephalon (a preparation called *cerveau isolè*) produced a bioelectrical activity that resembled sustained sleep. Although intense olfactory stimulation was able to provoke brief periods of wakefulness, which did not last beyond the duration of the stimulus, visual information could not 'wake up' the animal or elicit ocular signs or a diffuse EEG activation. Such experiments permitted identifying regions that participate in producing wakefulness, located in the mesencephalic reticular formation and in the posterior region of the hypothalamus. The difference between the 'cerveau isolè' preparation, which exhibits alternation of sleep and wakefulness, and the large proportion of wakefulness in animals with 'mediopontine' lesions indicates that a hypnogenic region responsible for EEG synchronization must be located in the posterior bulbopontine zone.

The demonstration of the existence of hypnogenic actions from structures lying in the anterior brain goes back to the pioneering experiments of Hess (1944). The electrical stimulation of thalamic regions elicits sleep in animals, which resembles the natural behaviour of the species.

By systemic injections of cholinergic drugs to pontine cats, Jouvet (1962) obtained an increase in the number of PS-like episodes. It has been demonstrated that direct application of acetylcholine crystals in the preoptic region caused sleep (Hernández-Peón et al., 1963). This effect could be blocked by applying atropine in more posterior brainstem regions, which suggested a rostrocaudal direction to the hypnogenic influence (Velluti and Hernández-Peón, 1963). Later, microinjection of acetylcholine or cholinergic agents in the brainstem (Cordeau et al., 1963) and pontine regions (George et al., 1964; see Gillin et al., 1985; Vivaldi et al., 1980; Baghdoyan et al., 1984; Reinoso-Suárez et al., 1994) evoked PS-like signs.

Wakefulness and sleep are characterized by different rhythms in the EEG and are associated with the changes in the cortical excitability that accompanies behaviour. The discharge of reticular thalamic neurons synchronize the forebrain activity during SWS, inducing the slow EEG oscillations at delta frequencies and the 12–16 cps frequencies typical of the spindles (Steriade, 2005).

The cerebral cortex, representing an 80% of the human CNS, must also participate in the generation and maintenance of sleep processes. Such aspects have not been sufficiently investigated so far.

Particular Processes of Paradoxical Sleep

Once in SWS, a prior and essential stage, the brain is able to produce the cortical electrographic activation and other PS signals. Two ascending inputs to the cortex are involved in this process; one originates in the intralaminar nuclei of the thalamus and another in the caudal hypothalamus.

Short and long PS episodes were described in rats as two populations with different functional meaning (Amici et al., 2005).

The cerebellar lesion is also capable of eliciting transitory changes in the sleep cycle. The unitary activity, both of Purkinje cells and of neurons from the fastigial nucleus, increases at the beginning of sleep, suggesting a probable cooperation between this nucleus and pontine zones of common embryologic origin (Mano, 1970; Hobson and McCarley, 1972). Moreover, the 'pO_2 PS system' includes cerebellar areas also supporting the notion of cerebellar participation in PS (Velluti, 1985).

During PS the spinal motor neurons are inhibited, which explains the atonia or hypotonia of certain muscles at this stage of sleep (Pompeiano, 1967; Lai and Siegel, 1991). Recent studies show that atonia may be produced by direct projections from the mesopontine tegmentum to the spinal cord (Lu et al., 2006).

The pons has a bioelectrical activity of its own that spreads throughout the brain, called ponto-geniculo-occipital waves (PGO; Fig. 2.12). This activity appears in a very showy manner in the transition between the SWS and PS, heralding and being part of a new PS episode (Jouvet et al., 1965; McCarley et al., 1978). In the cerebellar nuclei of the cat, PGO were also reported during spontaneous normal PS, under reserpine effects, and after microinjections of acetylcholine in the pons (Fig. 11.2; Velluti et al., 1985). Difficulties in recording similar waves in the lateral geniculate body of the albino rat led to some speculation that the fundamental PGO wave did not occur in this common laboratory animal, which tended to obscure what we believe to be a general phenomenon across mammalian species (Morrison, 2008).

At the cellular level was found that PGO waves in the lateral geniculate body of cats resulted from nicotinic activation of the projecting neurons with a parallel muscarinic inhibition of perigeniculate cells stemming from activation by pontine peribrachial cholinergic neurons. (Morrison, 2008).

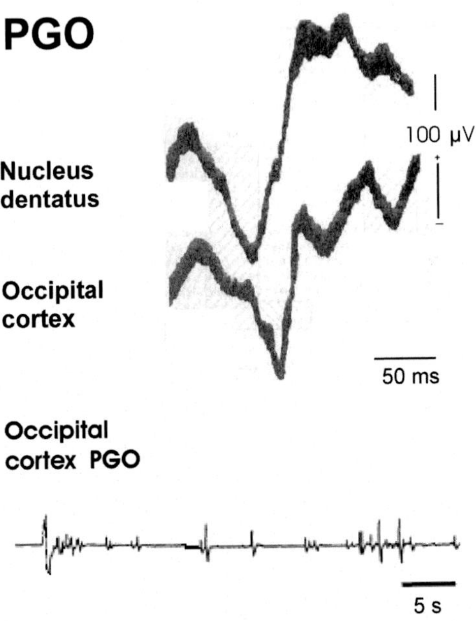

Figure 2.12 Example of PGO in the cerebellar nucleus *dentatus* and in the occipital cortex. PGO waves have also been found in the *fastigii* and interpositus cerebellar nuclei, with similar characteristics of amplitude and timing to PGO waves described in other areas *Modified from Velluti, R.A., Yamuy, J., Hadjez, J., Monti, J.M., 1985. Spontaneous cerebellar nuclei PGO-like waves in natural paradoxical sleep and under reserpine. Electroenceph. Clin. Neurophysiol. 60, 243–248.*

Ponto-geniculo-occipital Waves and Alerting

PGO waves, or spikes as they were originally named, are macropotential waves first recorded in the pontine tegmentum, lateral geniculate body, and the occipital (visual) cortex of cats, hence the acronym. The last two sites in particular reflected the early interest in dreams and eye movements among early sleep researchers. In cats, the waves are essentially limited to about 30 seconds preceding the onset of REM sleep and for the duration of that state. Only a few waves precede REM sleep in rats. There appear to be two types of PGO waves as they are recorded in cats: Those occurring singly and those appearing as bursts during an episode of REM sleep. The latter were eliminated following bilateral lesions in the vestibular nuclei that, which also blocked the occurrence of REMs (Morrison and Pompeiano, Arch Ital Biol. 1966), suggesting that the PGO bursts are reinforced by the nystagmoid REMs. Datta (Cell Mol Neurobiol. 1997) reviewed his laboratory's work that revealed interactions between the pontine generator site identified by him and the

(*Continued*)

(Continued)

vestibular nuclei, supporting the observations of Morrison and Pompeiano (1966). The single waves generated in the caudolateral parabrachial and subcoeruleus areas, then, appear to be a primary element of REM sleep, unlike the eye movement-associated bursts.

Difficulties in recording similar waves in the lateral geniculate body of the albino rat led to some speculation that the fundamental PGO wave did not occur in this common laboratory animal, which tended to obscure what we believe to be a general phenomenon across mammalian species (Kaufman and Morrison, Brain Res. 1981; Reiner and Morrison, Exp. Neurol. 1980). In fact, macropotential waves were later recorded in the dorsolateral pontine tegmentum that preceded the onset of REM sleep as in cats. Datta et al. (Synapse, 1998) demonstrated that the pontine generator cells do not project directly to the lateral geniculate body in the visually deficient albino rat, possibly explaining the earlier difficulties.

The focus on eye activity and dreams may have led workers away from what is in our opinion their real significance: a sign of alerting in the brain, a fundamental – and 'peculiar' – aspect of REM sleep. The idea that PGO waves are a sign of an 'alert' brain largely cut off from, or at least minimally responsive to, both external and internal influences arose when Bowker and Morrison (Brain Res. 1976) demonstrated in cats that auditory stimulation (90 dB level) would elicit PGO waves (PGOE) during REM sleep and even non-REM sleep. The same results were later obtained in rats.

At the cellular level, Hu et al. (Neurosci. 1989) found that PGO waves in the lateral geniculate body of cats resulted from nicotinic activation of the projecting neurons with a parallel muscarinic inhibition of perigeniculate cells stemming from activation by pontine peribrachial cholinergic neurons. They confirmed that auditory stimuli elicited PGO waves in the lateral geniculate body and concluded, in agreement with the earlier, that 'these signals are the central correlates of orienting reactions elicited by sensory stimuli during waking (the so-called eye movement potentials) and by internally generated drives during paradoxical sleep'.

Adrian R. Morrison
University of Pennsylvania, Philadelphia (2008)

A Partial Hypothesis on PS Fragmentation

Deviations from normal values of certain vital functions such as the heart and the respiratory rates could not be indefinitely prolonged, which becomes a limiting factor for the duration of PS periods. I propose that this particular architecture of sleep is based on individuals dividing the amount of PS necessary into shorter epochs. Internal security systems would make us return to homoeostatic controlled SWS or W, to enter a new PS episode after some time, and thus achieving its still enigmatic and apparently essential state (Velluti, 1988).

Although the various experimental approaches permitted to know the location of encephalic sites that generate *signs* of wakefulness or *signs* of the different sleep stages, it is essential to remark that are partial approximations to how the brain is functioning in order to develop a wakefulness–sleep cycle with all its physiological characteristics. *The brain as a whole is the responsibility of the cycle.*

Endogenous Humoral Factors

Throughout the history of sleep research, diverse factors of endocrine or biochemical origin have been postulated to explain this cycle. Pieron (1913) proposed the existence of substances that, generated during wakefulness, would be removed during the sleep. Investigations carried out in rabbits described a nonapeptide found in the cerebrospinal fluid (CSF) after electrical stimulation of the thalamus that induced sleep (Monnier and Hösli, 1964). Because injection into the brain ventricles of other rabbits provoked delta waves in the EEG, this peptide has been denominated *delta sleep-inducing peptide*. In most of the subsequent studies, this factor has showed only a mild hypnogenic effect. On the other hand, just like any other peptide, its normal passage through the hematoencephalic barrier is, in any case, difficult and slow.

Peptides derived from the propiomelanocortin and peptides immunologically active have also been proposed as hypnoinducers. More recently, two types of sleep-facilitator peptides have been proposed: (1) SWS promoting substances, such as growth hormone releasing hormone, interleukin-1β, tumour necrosis factor α, adenosine and prostagalndine D_2; and (2) PS-promoting substances, such as vasoactive intestinal polypeptide and prolactin. All these substances have fulfilled the criteria for sleep regulatory substances (Krueger and Obal, 1994; Obal and Krueger, 2005).

Hypnogenic actions are attributed nowadays to melatonin. Recent studies have demonstrated that this hormone, known for its chronobiotic effects for a long time, might also have an effect on sleep, causing it to decrease its latency and increase its efficiency (Cardinali, 2005).

The amount of substances that have been described in recent decades permit us to conclude that none of them has a powerful and determinant action, neither can they be attributed a sleep-provoking effect with physiological characteristics acting isolated. Perhaps the encompassed physiological actions of many of them generate a global modulation thus creating a situation favourable to sleep.

Neurotransmitters and Neuromodulators

The study of different neurotransmitters by administration of agonists and blockers has permitted inferring that some of them play a more central role than others in the generation or maintenance of the different stages of the sleep-wakefulness cycle. Considering that neurons receive hundreds of synapses and project to varied neuronal networks, whose function changes depending on the moment of the cycle, such neurochemical studies appear as simplistic approaches. They often lead to interpreting an effect as the result of the action of a certain neurotransmitter while it may simply be the result of an imbalance in a complex neural network. Bearing in mind such limitations, we can consider that certain neuronal systems and its neurotransmitters may act in certain moments of the cycle.

Among the systems involved in the generation and maintenance of wakefulness we can highlight the noradrenergic neurons in the *locus coeruleus*, the serotoninergic neurons in the dorsal *raphe* nuclei, the histaminergic neurons in the tuberomammillary nuclei, the hypocretin/orexinergic neurons in the perifornical area, the glutamatergic neurons in the reticular formation and the cholinergic neurons in the basal brain. These systems of neurons reinforce each other and operate in specific behaviours during wakefulness by converging on common effectors in the thalamus and the cortex (Brown and McCarley, 2005).

Neurons of the anterior hypothalamus that produce gamma amino butyric acid (GABA) are involved in SWS generation. Both GABA-enhancer and GABA-mimetic drugs increase the amount of SWS.

The ultradian cycle SWS-PS seems to be generated by the interaction of cholinergic and monoaminergic neurons (serotonin, noradrenalin) of the brainstem. This model indicates that, when the cholinergic neurons are released from the monoamine inhibition, they stimulate the neurons in the reticular formation, which in turn leads the signs of PS (see McCarley, 2004). However, selective lesions of either cholinergic or monoaminergic nuclei in the brainstem have limited effects on PS

(Jones et al., 1977; Shouse and Siegel, 1992). A recent report proposed a flip-flop switch model in the regulation of the switching into and out of PS (Lu et al., 2006). It was also suggested that additional circuitry must be involved in that switching, such as relevant descending pathways from the hypothalamus recently found in rat PS (Saper et al., 2005). Lateral hypothalamic hypocretinergic neurons may act on PS-off and PS-active brainstem units by means of such inhibitory transmitters as galanin and GABA.

Furthermore, the putative relevance of the greatest anatomical part of the primate brain, the neocortical mantle, should be recognized. The neocortex certainly influences or determines the brainstem and other sleep-related *loci* activity, which has been, so far, almost not experimented on.

Many neurotransmitters and hormones are involved in SWS control. After diverse lesion studies and precursor injections, serotonin is nowadays considered to act indirectly on sleep induction by modulating other hypnogenic factors of the anterior hypothalamus and the suprachiasmatic nucleus (Brown and McCarley, 2005).

Acetylcholine and Cholinergics

Acetylcholine microinjection techniques have demonstrated that the generation of PS-like signs is related to the activity of pontine regions (see Gillin et al., 1985; Vivaldi et al., 1980; Baghdoyan et al., 1984; Reinoso-Suárez et al., 1994). An outstanding fact is the latency of seconds or minutes between the acetylcholine injection and the beginning of the PS signs. The cholinergic substances appear to initiate a process that requires a certain time to develop or to get other regions started (Velluti et al., 1985; Velluti, 1988).

Noradrenaline might act together with acetylcholine during wakefulness, but would not operate during PS, since during this stage the noradrenergic neurons of the *locus coeruleus*, at least, remain silent.

The purine adenosine differs from all the neurotransmitters and neuromodulators described previously in several ways. It is present and released from many neurons, since it is a byproduct of cellular metabolism and is formed by the breakdown of the energy molecule ATP. Adenosine is not released from synaptic vesicles but instead is conveyed to the extracellular space by plasma membrane transporters. It has been implicated as a sleep-promoting factor (Porkka-Heiskanen et al., 1997; Basheer et al., 2000). In rats and cats in many areas of the brain adenosine levels rise during waking and drop substantially during SWS. In particular, in the basal forebrain adenosine levels rise during waking and, importantly, they continue to rise during the wakefulness induced by sleep deprivation. Moreover, adenosine inhibits identified cholinergic neurons in vitro, whereas in vivo, adenosine inhibits wake-active neurons (Brown and McCarley, 2005).

Studies have found the hypocretin/orexin neuronal neuromodulator system participating in the motor activity of both wakefulness and PS. This system would also be involved in certain pathologies related to sleep, such as the cataplexy of the narcolepsy (Taheri et al., 2002; Mignot, 2004).

> *'It is still possible that learning mechanisms are ascribed to the dynamic, emergent properties of neural ensembles. We have more neurons than proteins, and perhaps the former can carry out a good job without the need of any structural modifications of their already sophisticated connectivity. Why, then, do most neuroscientists prefer to lean on* ***neural plasticity*** *rather than on* ***neural functional states****? The most parsimonious answer is that we have collected a huge amount of information about the structure and connectivity of neural tissue at subcellular and molecular levels, and about the anatomical and biochemical rules and pathways maintaining these structures and circuits. In addition, definite behaviour and sensory-motor properties are easily ascribed to specific neural sites. In contrast, our information about brain functioning during learning situations is too constrained by the limitations imposed by electrical recordings from small numbers of neural elements selected out of billions, or by modern mapping techniques dealing with electrical or biochemical representations of brain activity.'*
>
> ***Jose Maria Delgado, 2008***

POSSIBLE FUNCTIONS PERFORMED DURING SLEEP

We know that body functions are influenced by the alternation of wakefulness and sleep; however, it is still unknown why we sleep. The most accepted hypotheses at present about the biological functions of sleep can be summarized as follows.

Recovery and Restoration

The hypothesis that the sleep serves the recovery and restoration of biochemical and physiological processes degraded during wakefulness is apparently logical and widely accepted. The increase in GH during SWS in humans would support this idea, although some other species do not exhibit such temporal correlation, such as rhesus monkeys and dogs. Whereas in humans there is a correlation between the duration of the preceding wakefulness and the subsequent sleep, this fact is not so clear in other species.

The effect of physical exercise on the subsequent sleep does not support the hypothesis of the restitution of the body in general. Contrary to this hypothesis is the fact that exercise carried out during the hours prior to sleep provokes delays in its installation and a lag of the circadian rhythm.

The CBF during SWS is always lower than during the corresponding PS, but the absolute value of SWS CBF 'at the beginning of the night can be higher than the PS value at the end of the night. Thus reduced SWS (0.25–4.0 cycles/s) and reduced metabolic activity towards the end of sleep suggest that some kind of recovery has occurred' (Zoccoli et al., 2005).

Energy Conservation

During SWS inactivity, the lower metabolic rate and body temperature reduces energy consumption. The metabolism reduction during sleep is of about 10% in relation to the basal levels of wakefulness. However, the energy conservation can be higher at low temperatures. The metabolic reduction in an unclothed human subject exposed to a room temperature of 21°C may reach 40% after sleep onset.

Plasticity, Memory and Learning

Sleep appears essential for cognitive development, although we are still far from knowing what processes are triggered during this period. However, memory and learning appear as relevant functions affected by sleep.

Recent data show that the learning process improves when followed by a night's sleep; it could be inferred that the SWS would be more closely related to the brain plasticity than to the total organism restoration (Cipolli, 2005). In addition, memory loss, learning difficulties, decrease in motor skills and mood changes are among the earliest signs of sleep deprivation.

Two general hypothesis of how sleep participates in memory consolidation have been proposed: (1) The *dual-process hypothesis* conceives that procedural memory benefits derive mainly from PS and that declarative memory is mainly consolidated during SWS II-IV stages. (2) The *sequential hypothesis* assumes that the occurrence of both SWS and PS is necessary to improve consolidation of adaptive and especially of procedural memories (Giuditta et al., 1995; Stickgold et al., 2000; Cipolli; 2005). Moreover, an explanatory hypothesis has been put forward that the consolidation of memories acquired in a previous waking is due to the offline reprocessing that they undergo during sleep (Stickgold, 1998).

Mismatch negativity (MMN) is an electrophysiological signal that can be correlated with learning, which changes during PS (Atienza et al., 2002). Furthermore, if a breast-fed infant is stimulated with a regular series of auditory stimuli and deviant stimuli randomly alternating during SWS —the deviant stimuli producing a MMN — this response is repeated with the same auditory series during the following wakefulness (Cheour et al., 2002). Thus newborns can assimilate auditory information during sleep. Furthermore, animal research shows that neurons, recorded from the primary auditory cortex of a guinea pig during SWS can discriminate a natural stimulus (conspecific vocalization) from its copy inverted in time (see Chapter 5: Auditory Unit Activity in Sleep). All of this makes the sleep as a whole essential for a normal cognitive development and organization of the information obtained during wakefulness.

As a first general conclusion it seems evident that the main objectives of the SWS and the PS remains unknown. It has been demonstrated that the ultimate general objective of sleep is not to provide a resting period to the CNS or the body. Moreover, although today it is not possible to establish the ultimate reason for sleep, according to what we have seen, the reasons must be highly diverse and certainly research cannot be postponed.

Finally, I introduce the idea of *psychophysiological information homoeostasis.* The learning processes, particularly those carried out during sleep, would decide what data to remember or forget in a moment of life or throughout life. Psychophysiological information homoeostasis maintains the memory load within a range that the brain needs to maintain in order to preserve the cognitive functions. To remember or to forget are part of our brain goals in order to embrace a healthy status.

REFERENCES

Adametz, J.H., 1959. Recovery of functions in cats with rostral lesions. J. Neurosurg. 16, 85–97.

Amici, R., Jones, C.A., Perez, E., Zamboni, G., 2005. A physiological view of REM sleep structure. In: Parmeggiani, P.L., Velluti, R.A. (Eds.), The Physiologic Nature of Sleep. Imperial College Press, London, pp. 161–185.

Aserinski, E., Kleitman, N., 1953. Regularly occurring periods of eye motility and concomitant phenomena during sleep. Science 118, 273–274.

Atienza, M., Cantero, J.L., Domínguez-Marín, E., 2002. Mismatch negativity (MMN): an objective measure of sensory memory and long-lasting memories during sleep. Int. J. Psychophysiol. 46, 215–225.

Baghdoyan, H.A., Rodrigo-Angulo, M.L., McCarley, R.W., Hobson, J.A., 1984. Site-specific enhancement and suppression of desynchronized sleep signs following cholinergic stimulation of three brainstem regions. Brain Res. 306, 39–52.

Basheer, R., Porkka-Heiskanen, T., Strecker, R.E., Thakkar, M.M., McCarley, R.W., 2000. Adenosine as a biological signal mediating sleepiness following prolonged wakefulness. Biol. Signals Recept. 9, 319–327.

Bastuji, H., García-Larrea, L., 1999. Evoked potentials as a tool for the investigation of human sleep. Sleep Med. Rev. 3, 23–45.

Bastuji, H., García-Larrea, L., 2005. Human Auditory Information Processing during Sleep. In: Parmeggiani, P.L., Velluti, R.A. (Eds.), The Physiologic Nature of Sleep. Imperial College Press, London, pp. 509–534.

Bernardi, G., Cecchetti, L., Francesca Siclari, F., Buchmann, A., Yu, X., Handjaras, G., et al., 2016. Sleep reverts changes in human gray and white matter caused by wake-dependent training. NeuroImage 367–377. Available from: http://dx.doi.org/10.1016/j.neuroimage.2016.01.020.

Brandenberger, G., 1993. Episodic hormone release in relation to REM sleep. J. Sleep Res. 2, 193–198.

Brandenberger, G., 2005. Endocrine correlates of sleep in humans. In: Parmeggiani, P.L., Velluti, R.A. (Eds.), The Physiologic Nature of Sleep. Imperial College Press, London, pp. 433–453.

Braun, A.R., Balkin, T.J., Wesensten, N.J., Carson, R.E., Varga, M., Baldwin, P., et al., 1997. Regional cerebral blood flow throughout the sleep-wake cycle. An H2(15)O PET study. Brain 120, 1173–1197.

Bremer, F., 1935. Cerveau "isole" et physiologie du sommeil. C.R. Soc. Biol. 118, 1235–1241.

Brown, R.E., McCarley, R.W., 2005. Neurotransmitters, neuromudulators, and sleep. In: Parmeggiani, P.L., Velluti, R.A. (Eds.), The Physiologic Nature of Sleep. Imperial College Press, London, pp. 45–75.

Cardinali, D.P., 2005. The use of melatonin as a chronobiotic-cytoprotective agent in sleep disorders. In: Parmeggiani, P.L., Velluti, R.A. (Eds.), The Physiologic Nature of Sleep. Imperial College Press, London, pp. 455–488.

Cartwright, R.D., 1974. The influence of a concious wish on dreams: a methodological study of dream meaning and function. J. Abnorm. Psychol. 83, 387–393.

Cannon, W.B., 1929. Organization for physiological homeostasis. Physiol. Rev. 9, 399–431.

Campbell, K., Bell, I., Bastien, C., 1992. Evoked potential measures of information processing during natural sleep. In: Broughton, R.J., Ogilvie, R.D. (Eds.), Sleep, Arousal, and Performance. Birkhauser, Boston-Basel-Berlin, pp. 89–116.

Cespuglio, R., Colas, D., Gautier-Sauvigné, S., 2005. Energy processes underlying the sleep-wake cycle. In: Parmeggiani, P.L., Velluti, R.A. (Eds.), The Physiologic Nature of Sleep. Imperial College Press, London, pp. 3–21.

Chemelli, R.M., Willie, J.T., Sinton, C.M., Elmquist, J.K., Scammell, T., Lee, C., et al., 1999. Narcolepsy in orexin knockout mice: molecular genetics of sleep regulation. Cell 98, 437–451.

Cheour, M., Martynova, O., Naatanen, R., Erkkola, R., Sillanpaa, M., Kero, P., et al., 2002. Speech sounds learned by sleeping newborns. Nature 415, 599–600.

Chow, K.L., Randall, W., 1964. Learning and retention in cats with lesions in reticular formation. Psychon. Sci. 1, 259–260.
Cicogna, P., Cavallero, C., Bosinelli, M., 1991. Cognitive aspects of mental activity during sleep. Am. J. Psychol. 104, 413–425.
Cirelli, C., 2009. The genetic and molecular regulation of sleep: from fruit flies to humans. Nat. Rev. Neurosci. 10, 549–560.
Cipolli, C., 2005. Sleep and memory. In: Parmeggiani, P.L., Velluti, R.A. (Eds.), The Physiologic Nature of Sleep. Imperial College Press, London, pp. 601–629.
Coenen, A.M., 1995. Neuronal activities underlying the electroencephalogram and evoked potentials of sleeping and waking: implications for information processing. Neurosci. Biobehav. Rev. 19, 447–463.
Cordeau, J.P., Moreau, A., Beaulnes, A., Laurin, C., 1963. EEG and behavioral changes following micro-injections of acetylcholine and adrenaline in the brainstem of the cats. Arch. Ital. Biol. 101 (30), 47.
Dauvilliers, Y., Maret, S., Tafti, M., 2005. Genetics of normal and pathological sleep in humans. Sleep Med. Rev. 9, 91–100.
Dang-Vu, T.T., Manuel Schabus, M., Martin Desseilles, M., Sterpenich, V., Maxime Bonjean, M., Maquet, P., 2010. Functional neuroimaging insights into the physiology of human sleep. Sleep 33 (12).
Delgado, J.M., 2008. In: Velluti, R., The Auditory System in Sleep. pp. 135–136.
Durmer, J.S., Dinges, D.F., 2005. Neurocognitive Consequences of Sleep Deprivation. In: Roos, K.L. (Ed.), Sleep in Neurological Practice. Seminars in Neurobiology, vol. 25, pp. 117–129.
Esteban, S., Nicolau, M.C., Gamundi, A., Akaarir, M., Rial, R., 2005. Animal sleep: philogenetic correlations. In: Parmeggiani, P.L., Velluti, R.A. (Eds.), The Physiologic Nature of Sleep. Imperial College Press, London, pp. 207–245.
Evarts, E.V., 1964. Temporal pattern of discharge of pyramidal tract neurons during sleep and waking in the monkey. J. Neurophysiol. 27, 152–171.
Evarts, E.V., Bental, E., Bihari, B., Huttenlocher, P.R., 1962. Spontaneous discharge of single neurons during sleep and waking. Science 135, 726–728.
Foulkes, W.D., 1962. Dream reports from different stages of sleep. J. Abnorm. Soc. Psychol. 65, 14–25.
Franzini, C., 1992. Brain metabolism and blood flow duirng sleep. J. Sleep Res. 1, 3–16.
García-Austt, E., Velluti, R.A., Villar, J.I., 1968. Changes in brain pO_2 during paradoxical sleep in cats. Physiol. Behav. 3, 477–485.
George, R., Haslett, W.L., Jenden, D.J., 1964. A cholinergic mechanism in the brainstem reticular formation: induction of paradoxical sleep. Int. J. Neuropharmacol. 3, 541–552.
Gillin, J.C., Sitaram, N., Janowsky, D., Risch, C., Huey, L., Storch, F.I., 1985. Cholinetrgic mechanisms in REM sleep. In: Wauquier, A., Gaillard, J.M., Monti, J. M., Radulovacki, M. (Eds.), Neurotransmitters and Neuromodulators. Raven Press, New York, pp. 153–164.
Giuditta, A., Ambrosini, M.V., Montagnese, P., Mandile, P., Cotrugno, M., Grassi, Z.G., et al., 1995. The sequential hypothesis of the function of sleep. Behav. Brain Res. 69, 157–166.
Guilleminault, C., Anagnos, A., 2005. In: Kryger, M.H., Roth, T., Dement, W.C. (Eds.), Principles and Practice of Sleep Medicine. Elsevier Saunders, Philadelphia, p. 676.
Hernández-Peón, R., Chávez Ibarra, G., Morgane, J.P., Timo Iaria, C., 1963. Limbic cholinergic pathways involved in sep and emotional behavior. Exp. Neurol. 8, 93–111.
Hess, W.R., 1944. Das Schlafsyndrom als folge dienzephaler reizung. Helv. Physiol. Pharamcol. Acta 2, 305–344.

Hobson, J.A., McCarley, R.W., 1972. Spontaneous discharge rates of cat cerebellar Purkinje cells. Electroenceph. Clin. Neurophysiol. 33, 457–459.

Hobson, J.A., Pace-Schott, E.F., Stickgold, R., Kahn, D., 1998. To dream or not to dream? Relevant data from new neuroimaging and electrophysiological studies. Curr. Opin. Neurobiol. 8, 239–244.

Issa, E.B., Wang, X., 2008. Sensory responses during sleep in primate primary and secondary auditory cortex. J. Neurosci. 28 (53), 14467–14480.

John, E.R., 2001. The neurophysics of conciousness. Brain Res. Rev. 39, 1–28.

John, E.R., 2006. The sometimes pernicious role of theory in science. Int. J. Psychophysiol. 62, 377–383.

John, E.R., Ranshoff, J., 1996. Coordination of International Coma Treatment Consortium for NICHD.

Jones, B.E., Harper, S.T., Halaris, A.E., 1977. Effects of locus coeruleus lesions upon cerebral monoamine content, sleep-wakefulness states and the response to amphetamine in the cats. Brain Res. 124, 473–496.

Jouvet, M., 1962. Recherches sur les structures nerveuses et le mecanismes responsables de differentes phases du sommeil physiologique. Arch. Ital. Biol. 100, 125–206.

Jouvet, M., 1999. The paradox of sleep: the story of dreaming. The MIT Press, Boston, MA, USA.

Jouvet, M., Jeannerod, M., Delorme, F., 1965. Organisation du systéme responsable de l' activité phasique au course du sommeil paradoxal. C.R. Seanc. Soc. Biol. (Paris) 159, 1599–1604.

Kleitman, N., 1963. Sleep and Wakefulness. University of Chicago Press, Chicago-London.

Krueger, J.M., Obal Jr., F., 1994. Sleep factors. In: Saunders, N.A., Sullivan, C.E. (Eds.), Sleep and Breathing. Marcel Dekker Inc., New York, pp. 79–112.

Lai, Y.Y., Siegel, J.M., 1991. Pontomedullary glutamate receptors mediating locomotion and muscle tone suppression. J. Neurosci. 11, 2931–2937.

Lee, A.K., Wilson, M.A., 2002. Memory of sequential experience in the hippocampus during slow wave sleep. Neuron 36, 1183–1194.

Libert, J.-P., Bach, V., 2005. Thermoregulation and sleep in the human. In: Parmeggiani, P.L., Velluti, R.A. (Eds.), The Physiologic Nature of Sleep. Imperial College Press, London, pp. 407–431.

Loomis, A.L., Harvey, E.N., Hobart, G.A., 1938. Distribution of disturbance-patterns in the human electroencephalogram, with special reference to sleep. J. Neurophysiol. 1, 413–430.

Louie, K., Wilson, M.A., 2001. Temporally structured replay of awake hippocampal ensemble activity during rapid eye movement sleep. Neuron 29, 145–156.

Lu, J., Sherman, D., Devor, M., Saper, C.B., 2006. A putative flip-flop switch for control of REM sleep. Nature 441/1, 589–594.

Mano, N.I., 1970. Changes of simple and complex spike activity of cerebellar Purkinje cells with sleep and waking. Science 170, 1325–1327.

Maquet, P., 2000. Functional neuroimaging of normal sleep by positron emission tomography. J. Sleep Res. 9, 207–231.

Maquet, P., Dive, D., Salmon, E., Sadzot, B., Franco, G., Poirrier, R., et al., 1990. Cerebral glucose utilization during sleep—wake cycle in man determined by positron emission tomography and [18F]2-fl uoro-2-deoxy-d-glucose method. Brain Res. 513, 136–143.

Maquet, P., Degueldre, C., Delfi ore, G., Aerts, J., Peters, J.M., Luxen, A., et al., 1997. Functional neuroanatomy of human slow wave sleep. J. Neurosci. 17, 2807–2812.

Maquet, P.A.A., Sterpenich, V., Albouy, G., Dang-vu, T., Desseilles, M., Boly, M., et al., 2005. Brain imaging on passing to sleep. In: Parmeggiani, P.L., Velluti, R.A. (Eds.), The Physiologic Nature of Sleep. Imperial College Press, London, pp. 489–508.

Martinez-Orozco, F.J., Vicario, J.L., De Andresc, C., Fernandez-Arquero, M., Peraita-Adrados, R., 2016. Comorbidity of narcolepsy (type 1) with autoimmune diseases and other immunopathological disorders. J. Clin. Med. Res. 8 (7), 495–505.

McCarley, R.W., 2004. Mechanisms and models of REM sleep control. Arch. Ital. Biol. 142, 429–467.

McCarley, R.W., Nelson, J.P., Hobson, J.A., 1978. Ponto geniculo occipital (PGO) burst neurons: correlative evidence for neuronal generators of PGO waves. Science 20, 269–272.

McCarley, R.W., Hoffman, E.A., 1981. REM sleep dreams and the activation-synthesis hypothesis. Am. J. Psychiatry 38, 904–912.

McGinty, D., Szymusiak, R., 2003. Hypothalamic regulation of sleepand arousal. Front. Biosci. 960, 165–173.

Mignot, E., 2004. Sleep, sleep disorders and hypocretin (orexin). Sleep Med. 5 (Suppl. 1), S2–S8.

Mizraji, E., 2008. Cell assemblies and neural networks. In: Velluti, R.A. (Ed.), The Auditory System on Sleep. Academic Press, Elsevier, pp. 81–82.

Monnier, M., Hösli, L., 1964. Dialysis of sleep and waking factor in blood of the rabbit. Science 146, 796–798.

Morrison, A.R., 2008. Ponto-geniculo-occipital waves and alerting. In: Velluti, R.A. (Ed.), The Auditory System in Sleep. Elsevier, pp. 31–32.

Moruzzi, G., 1963. Active processes in the brain stem during sleep. Harvey Lect. Ser. 58, 253–297.

Moruzzi, G., Magoun, H., 1949. Brain stem reticular formation and activation of the EEG. Electroenceph. Clin. Neurophysiol. 1, 455–473.

Obal Jr, F., Krueger, J.M., 2005. Humoral mechanisms of sleep. In: Parmeggiani, P.L., Velluti, R.A. (Eds.), The Physiologic Nature of Sleep. Imperial College Press, London, pp. 23–43.

Orem, J., 1989. Behavioral inspiratory inhibition: Inactivated and activated respiratory cells. J. Neurophysiol. 62, 1069–1078.

Orem, J., Lovering, A.T., Dunin-Barkowski, W., Vidruk, E.H., 2002. Tonic activity in the respiratory system in wakefulness, NREM and REM sleep. Sleep 25, 488–496.

Orem, J.M., 2005. Neural control of breathing in sleep. In: Parmeggiani, P.L., Velluti, R. A. (Eds.), The Physiologic Nature of Sleep. Imperial College Press, London.

Parmeggiani, P.L., 1980. Temperature regulation during sleep: a study in homeostasis. In: Orem, H., Barnes, Ch. (Eds.), Physiology in Sleep. Academic Press, New York, pp. 98–136.

Parmegianni, P.L., 2011. Systemic Homeostasis and Poiquilostasis in Sleep. Imperial College Press, London.

Parmeggiani, P.L., 2005a. Sleep behaviour and temperature. In: Parmeggiani, P.L., Velluti, R.A. (Eds.), The Physiologic Nature of Sleep. Imperial College Press, London, pp. 387–406.

Parmeggiani, P.L., 2005b. Physiologic regulation in sleep. In: Kryger, M.H., Roth, T., Dement, W.C. (Eds.), Principles and Practice of Sleep Medicine. Elsevier Saunders, Philadelphia, pp. 169–178.

Parmeggiani, P.L., Sabattini, L., 1972. Electromyographic aspects of postural, respiratory and thermoregulatory mechanisms in sleeping cats. Electroenceph. Clin. Neurophysiol. 33, 1–13.

Pedemonte, M., Velluti, R.A., 2005a. What individual neurons tell us about encoding and sensory processing in sleep. In: Parmeggiani, P.L., Velluti, R.A. (Eds.), The Physiologic Nature of Sleep. Imperial College Press, London, pp. 489–508.

Pedemonte, M., Velluti, R.A., 2005b. Sleep hippocampal theta rhythm and sensory processing. In: Lander, M., Cardinali, D.P., Perumal, P. (Eds.), Sleep and Sleep Disorders: A Neuropsychopharmacological Approach. Landes Biosciences/Springer, TX/NY, USA.

Pedemonte, M., Peña, J.L., Torterolo, P., Velluti, R.A., 1996. Auditory deprivation modifies sleep in the guinea pig. Neurosci. Lett. 223, 1–4.

Peña, J.L., Pérez-Perera, L., Bouvier, M., Velluti, R.A., 1999. Sleep and wakefulness modulation of the neuronal fi ring in the auditory cortex of the guinea-pig. Brain Res. 816, 463–470.

Pieron, H., 1913. Le problème physiologique du sommeil. Masson & Cie, Paris.

Pompeiano, O., 1967. The neurophysiological mechanisms of the postural and motor events during desynchronized sleep. Proc. Assoc. Res. Nerve. Ment. Dis. 45, 351–423.

Porkka-Heiskanen, T., Strecker, R.E., Thakkar, M., Bjorkum, A.A., Greene, R.W., McCarley, R.W., 1997. Adenosine: a mediator of the sleep-inducing effects of prolonged wakefulness. Science 276, 1265–1268.

Portas, Ch.M., 2005. Cognitive aspects of sleep: perception, mentation, and dreaming. In: Parmeggiani, P.L., Velluti, R.A. (Eds.), The Physiologic Nature of Sleep. Imperial College Press, London, pp. 535–569.

Reinoso-Suárez, F., De Andrés, I., Rodrigo-Angulo, M.L., Rodríguez-Veiga, E., 1994. Location and anatomical connections of a paradoxical sleep induction site in the cat ventral pontine tegmentum. Eur. J. Neurosci. 6, 1829–1836.

Reivich, M., 1974. Blood flow metabolism couple in brain. In: Plum, F. (Ed.), Brain Dysfunction in Metabolic Disorders. Raven, N.Y, pp. 125–140.

Saper, C., B., Scamelli, T.E., Lu, J., 2005. Hypothalamic regulation of sleep and circadian rhythms. Nature 437, 1257–1263.

Sassin, J.F., Parker, D.C., Mace, J.W., Grotlin, R.W., Johnson, L.C., Rossman, L.G., 1969. Human growth hormone release: relation to slow wave sleep and sleep–waking cycles. Science 165, 513–515.

Sastre, J., Jouvet, M., 1969. Le comportement onirique du chat. Physiol. Behav. 22, 979–989.

Shouse, M.N., Siegel, J.M., 1992. Pontine regulation of REM sleep components in cats: integrity of the pedunculopontine tegmentum (PPT) is important for phasic events but unnecessary for atoinia during REM sleep. Brain Res. 571, 50–63.

Siclari, F., et al., 2017. The neural correlates of dreaming. Nat. Neurosci. Available from: http://dx.doi.org/10.1038/nn.4545.

Silvani, A., Lenzi, P., 2005. Reflex cardiovascular control in sleep. In: Parmeggiani, P.L., Velluti, R.A. (Eds.), The Physiologic Nature of Sleep. Imperial College Press, London, pp. 323–349.

Soca, F., 1900. Sur un cas de sommeil prolongé pendant sept mois par un tumeur de l'hypophyse. Nouv. Iconogr. Salpêtriére 2, 101–115.

Steriade, M., 2005. Brain electrical activity and sensory processing during waking and sleep states. In: Kryger, M.H., Roth, T., Dement, W.C. (Eds.), Principles and Practice of Sleep Medicine. Saunders Co, Philadelphia, USA.

Stickgold, R., 1998. Sleep of-line memory reprocessing. Trends Cogn. Sci. 2, 484–492.

Stickgold, R., Malia, A., Maguire, D., Roddenberry, D., O'Connor, M., 2000. Replaying the game: hypnagogic images in normals and amnesics. Science 290, 350–353.

Suntsova, N., Szymusiak, R., Alam, Md.N., Guzmán-Marín, R., McGinty, D., 2002. Sleep-waking discharge patterns of medial preoptic nucleus neurons in rats. J. Physiol. 543, 665–677.

Tafti, M., 2003. Deficiency in short chain fatty acid beta-oxidation affects theta oscillations during sleep. Nat. Genet. 34, 320–325.

Taheri, S., Zeitzer, J.M., Mignot, E., 2002. The role of hypocretins (orexins) in sleep regulation and narcolepsy. Ann. Rev. Neurosci. 25, 283–313.

Takahashi, Y., Kipnis, D.M., Daughaday, W.H., 1968. Growth hormone secretion during sleep. J. Clin. Invest. 47, 2079–2090.

Tononi, G., Cirelli, C., 2003. Sleep and synaptic homeostasis: a hypothesis. Brain Res. Bull. 62, 143–150.

Tononi, G., Cirelli, C., 2005. A possible rol for sleep in synaptic homeostasis. In: Parmeggiani, P.L., Velluti, R.A. (Eds.), The Physiologic Nature of Sleep. Imperial College Press, London, pp. 77–101.

Velluti, R., Hernández-Peón, R., 1963. Atropine blockade within a cholinergic hypnogenic circuit. Exp. Neurol. 8, 20–29.

Velluti, R., Monti, J.M., 1976. pO_2 recorded in the amygdaloid complex during the sleep-waking cycle in cats. Exp. Neurol. 50, 798–805.

Velluti, R., Roig, J.A., Escarcena, L.A., Villar, J.I., García-Austt, E., 1965. Changes of brain pO_2 during arousal and alertness in unrestrained cats. Acta Neurol. Latinoam. 11, 368–382.

Velluti, R., Velluti, J.C., García-Austt, E., 1977. Cerebellum pO_2 and the sleep waking cycle in cats. Physiol. Behav. 18, 19–23.

Velluti, R.A., 1985. An electrochemical approach to sleep metabolism: a pO_2 paradoxical sleep system. Physiol. Behav. 34, 355–358.

Velluti, R.A., 1988. A functional viewpoint on paradoxical sleep-related brain regions. Acta Physiol. Pharmacol. Latinoam. 38, 99–115.

Velluti, R.A., 1997. Interactions between sleep and sensory physiology. J. Sleep Res 6, 61–77.

Velluti, R.A., 2005. Remarks on sensory neurophysiological mechanisms participating in active sleep processes. In: Parmeggiani, P.L., Velluti, R.A. (Eds.), The Physiologic Nature of Sleep. Imperial College Press, London, pp. 247–265.

Velluti, R.A., Pedemonte, M., 2002. *In vivo* approach to the cellular mechanisms for sensory processing in sleep and wakefulness. Cell. Mol. Neurobiol. 22, 501–515.

Velluti, R.A., Pedemonte, M., 2012. Sensory neurophysiologic functions participating in active sleep processes. Sleep Sci. 5 (4), 103–106.

Velluti, R.A., Peña, J.L., Pedemonte, M., 2000. Reciprocal actions between sensory signals and sleep. Biol. Signals Recept. 9, 297–308.

Velluti, R.A., Yamuy, J., Hadjez, J., Monti, J.M., 1985. Spontaneous cerebellar nuclei PGO-like waves in natural paradoxical sleep and under reserpine. Electroenceph. Clin. Neurophysiol. 60, 243–248.

Vivaldi, E., McCarley, R.W., Hobson, J.A., 1980. Evocation of desynchronized sleep signs by chemical microstimulation of the pontine brain stem. In: Hobson, J.A., Brazier, M.A.B. (Eds.), The Reticular Formation Revisited. Raven Press, New York, pp. 513–529.

von Economo, C., 1930. Sleep as a problem of localization. J. Nerv. Ment. Dis. 7, 249–259.

Yamamoto, T., Lyeth, B.G., Dixon, E.D., Robinson, S.E., Jenkins, L.W., Young, H.E., et al., 1988. Changes in regional brain acetylcholine content in rats following unilateral and bilateral brainstem lesions. J. Neurotrauma 5, 69–79.

Zoccoli, G., Bojic, T., Franzini, C., 2005. Regulation of cerebral circulation during sleep. In: Parmeggiani, PL, Velluti, RA (Eds.), The Physiologic Nature of Sleep. Imperial College Press, London, pp. 351–369.

FURTHER READING

Atienza, M., Cantero, J.L., 1997. The mismatch negativity component reveals the sensory memory memory during REM sleep in humans. Neurosci. Lett. 237, 21–24.

Kakigi, R., Naka, D., Okusa, T., Wang, X., Inui, K., Qiu, Y., et al., 2003. Sensory perception during sleep in humans: a magnetoencephalograhic study. Sleep Med. 4, 493–507.

Portas, Ch.M., Krakow, K., Allen, P., Josephs, O., Armony, J.L., Frith, C.D., 2000. Auditory processing across the sleep-wake cycle: simultaneous EEG and fMRI monitoring in humans. Neuron 28, 991–999.

Roffwarg, H.P., Muzio, J.N., Dement, W.C., 1966. Ontogenetic development of the human sleep-dream cycle. Science 152, 604–619.

Tononi, G., Cirelli, C., 2014. Sleep and the price of plasticity: from synaptic and cellular homeostasis to memory consolidation and integration. Neuron 81, 12–34.

Velluti, R.A., Pedemonte, M., García-Austt, E., 1989. Correlative changes of auditory nerve and microphonic potentials throughout sleep. Hear. Res. 39, 203–208.

CHAPTER 3

Notes on Information Processing

In general terms, information processing can be defined as a shift of information in some manner detectable by an observer or a system. In relation to biological sensory systems, it may be a process that describes all that happens in the outer world or in the body, such as the nature sudden sounds, the change in heart rate when rapidly awakened or after running, or a headache.

This processing has more specifically been defined by Shannon (1948) as the conversion of latent information into manifest information: 'The fundamental problem of communication is that of reproducing at one point, either exactly or approximately, a message selected at another point' (Fig. 3.1). From a cognitive approach, information processing is an instrument to try to reach an understanding of human memory storage or learning processes. Therefore, equivalent objectives can be followed in sleep as well.

Information processing can be sequential or in parallel, both of which can be either centralized or distributed. The parallel distributed processing has lately become known under the name of connectionism. The idea of spontaneous order in the brain arising out of decentralized networks of simple units, neurons, was already put forward in the 1950s.

The world is a highly structured place. This structure is reflected by the fact that signals that reach our sense organs are not completely at random but rather exhibit correlations in space and time. The receptors, pathway nuclei and cortical loci process and distribute the space and time information; thus, the brain must deal with both place neurons (space-related units represented in the hippocampus) and our theta rhythm phase-locked auditory units, that is, time correlated (Liberman et al., 2009; see Chapter 5, Auditory Unit Activity in Sleep).

Information entropy is the average number of bits needed for data storage or communication. Experiments in several systems demonstrate that real neurons and synapses approach the limits to information transmission set by the spike train or synaptic vesicle entropy. Rather than throwing away information in favour of specific biologically relevant

The Auditory System in Sleep
DOI: https://doi.org/10.1016/B978-0-12-810476-7.00003-8

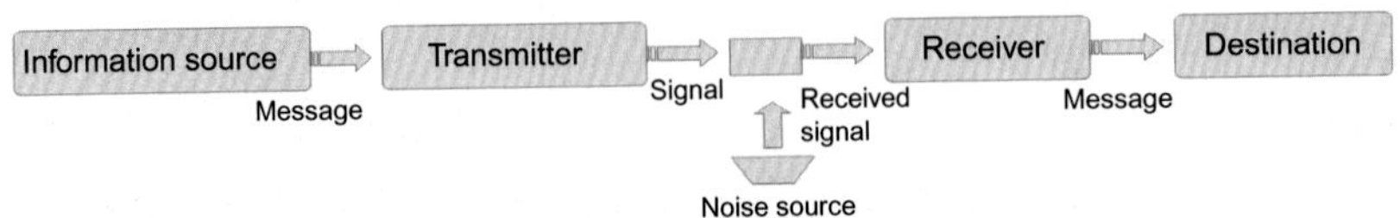

Figure 3.1 A communication system according to Shannon. The information source selects a message; the transmitter changes the message into a set of signals, which are in turn sent to the receiver. The receiver converts the signal back into the message. Noise is added to the signal. *Modified from Shannon, C.E. 1948. A mathematical theory of communication. Bell Sys. Tech. J. 27: 379–423. Reused with permission of Nokia Corporation.*

signals, these neurons seem to pack as much information as possible into the spike sequences they send to the brain, so that the same number of spikes can be used to transmit more information about the most structured signals coming from the real world (Rieke et al., 1997).

Cell assemblies and neural networks

The concept of cell assembly has been one of the bases for the development of mathematical models of distributed memories, an enterprise performed by various researchers around 1970. These models showed how information encoded in thousands of parallel firing neurons can be stored into a large neuronal network. The mathematical representation of the global activity of a large cell assembly is a large dimensional numerical vector (Mizraji and Lin, 2017). This vector is composed by numbers that measure the biophysical activities of the neurons (e.g., firing rates). Some memory models store pairs of input–output vectors. The material counterpart of these models suggests that the physical residences of memory traces are the synapses (Arbib, 2002).

A consequence of these models was the creation of many biologically inspired learning algorithms. These algorithms are mathematical procedures that usually modify the synaptic strengths in response to the intensity of the local synaptic inputs and the global neuronal output. Among the powerful learning rules, let us mention the Widrow–Hoff algorithm for single neuronal layer memories, and the backpropagation algorithm for multilayer memories (or memories with "hidden layers"). The distributed memories models have interesting "biomimetic" properties. For instance, in large networks partial damages of the physical support of the memories (e.g., destructions of neurones or synapses) do not produce catastrophic deteriorations of the stored data. Other example of biomimesis is the following: Some distributed memories models are capable of building prototypes from a partial exposition to data, a property that emulates the cognitive construction of general concepts from partial experiences; these prototypes are statistical averages emerged

(*Continued*)

(Continued)

from the experienced perceptual inputs and are represented by large dimensional vectors.

The neural representation of the human language is an ambitious (and yet far) objective of the neural network theory. Nevertheless, some important advances have occurred in recent times. In this sense, we mention the model proposed by Elman to represent the language processing as a dynamical time-depending activity. This model uses a set of modular neural networks that performs the conversion of phonetic inputs into conceptual outputs, and includes an associative memory and a working memory. The working memory produces contexts that help to decode the phonetic inputs entering the conceptual memory during the subsequent time steps. The original Elman's model uses associative memories containing hidden layers, and a miniature language is installed in the memory using the backpropagation learning algorithm. This model has been adapted by Hoffman and McGlashan to explore aspects of the neurodevelopmental hypothesis about the origin of schizophrenic disorders. In particular, these authors focused on the emergence of verbal hallucinations, and created a computational protocol to numerically simulate the generation of hallucinated voices in neural networks with exacerbated synaptic disconnections. Hoffman–McGlashan approach can be expanded in various directions, and the approach can be implemented using different neural models.

All these models allow to experiment "in numero" the effect of different pharmacological (together with psychological) therapies on the hallucinations. Neural networks models offer to medical research a promising integrative instrument to help with experimental and clinical research (Spitzer, 1999; Hoffman et al., 2011).

Eduardo Mizraji

Facultad de Ciencias, Universidad de la República, Montevideo, Uruguay

CODING

For any kind of neural processing, and particularly for auditory data processing, a basic code is needed. In general, it is likely that this basic code is that of a cell assembly, that is, the assembled activity of a population of neurons (Sakurai, 1999). Understanding the neural coding is significant to understanding the relation between the events in the sensory incoming data from the world or body and the spike trains. We may assume that the code is similar during the dissimilar brain states considered, wakefulness and sleep stages. The differences between states would then be based on the configuration of the new organized neuronal network/cell assembly active regions. The final result of the processing during wakefulness or sleep will be different, taking into account that the brain regions involved have changed. That is, the brain is, on a large scale, functionally different upon passing to sleep phases.

A dynamic modulation of the correlated neuronal firing may occur; that is, an auditory neuron that participates in a certain functional cell assembly/network may later on become associated with another activated and perhaps competing neuronal assembly on passing from wakefulness into sleep. The partial overlapping of neurons among assemblies is due to the ability of one neuron to participate in different types of information processing, based on the enormous amount of synapses each neuron may receive. Moreover, this condition may be repeated during the many and diverse waking states, during human sleep stage I or II and slow wave sleep, and also during phasic or tonic epochs of paradoxical sleep.

It is usually assumed that information is carried by the firing rate. Modulation of many biological sounds (frog call, guinea pig call, speech, etc.) occurs in a very short time, 5–20 ms. During this time, a neuron that normally fires, say, 100 spikes per second, can generate just 1 or 2 spikes. A 'timing code' includes both the spike count and the time quantity measuring the probability of spike occurrence. Barn owls can localize a prey by acoustic cues alone, with an auditory system that can recognize different times of signal (spike) arrival as low as 1 ms. There is no question that this temporal information is essential to such a discrimination task. It has been identified that the neural circuit responsible for the precise temporal measurement of phase-locked spikes coming from the two ears (Carr and Konishi, 1990). Moreover, different temporal spike patterns in the presynaptic nerve fibres may result in a completely different postsynaptic response. Two different patterns of presynaptic experimental pulses demonstrated that the presynaptic timing is important in determining the postsynaptic response (Segundo et al., 1963).

NEURONAL NETWORK/CELL ASSEMBLY

The concept of neuronal assemblies is defined by the temporally correlated neuronal firing associated to some functional aim and the most likely information coding is the ensemble coding by cell assemblies. Neuronal groups connected with several other neurons or groups can carry out functional cooperation and integration among widely distributed cells even with different functional properties to subserve a new state or condition. On the other hand, an individual neuron receives thousands of synaptic contacts on its membrane that turn its activity into a continuous membrane potential fluctuation, which determines a very unstable physiological condition to constitute a basic code for information processing. Furthermore, the neuronal network/cell assembly may provide a selective synaptic activity enhancement referring to a dynamic and transient efficacy which I suggest to be correlated to the behavioural dynamic modulation of the sleep process. That is, a neuron firing in a functional associated group may process some information and, some time later, may become associated with other competing and activated neuronal groups for different functional purposes, such as after passing from wakefulness into sleep.

These diverse and new neuronal associations may occur during the wakefulness states, during slow wave sleep (III–IV), in human sleep stage I or II and during paradoxical sleep, phasic or tonic epochs.

Fig. 3.2 explains the basic possibilities or properties of a cell assembly coding. Schematically, it shows a partial overlapping of neurons. Some of them belong to two different neuronal networks, while a second physiological possibility is the switch of others from one state to another, that is, construction and reconstruction of assemblies.

Progressing with this analysis, let us compare what happens in the anatomofunctional 'brain' networks of a simple mollusc, *Tritonia*, capable of two different motor abilities (Fig. 3.3), which, extrapolating, could represent two different basic components in a complex brain as the waking and sleep states. Getting (1989), postulated that 'If these network, synaptic, and cellular mechanisms are under modulatory control, then an

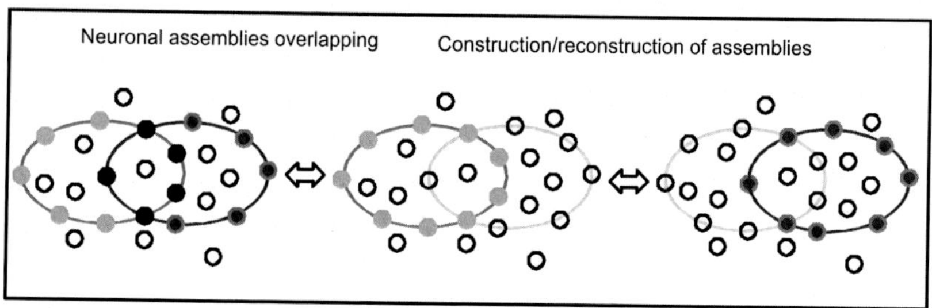

Figure 3.2 Examples of the manifold possible encoding properties and neuronal network/cell assembly combinations of active neurons. The arrows indicate possible and minimal dynamics of constructions and reconstructions of cell assemblies. This is an oversimplified approach of what can occur throughout the brain during the sleep-waking shift.

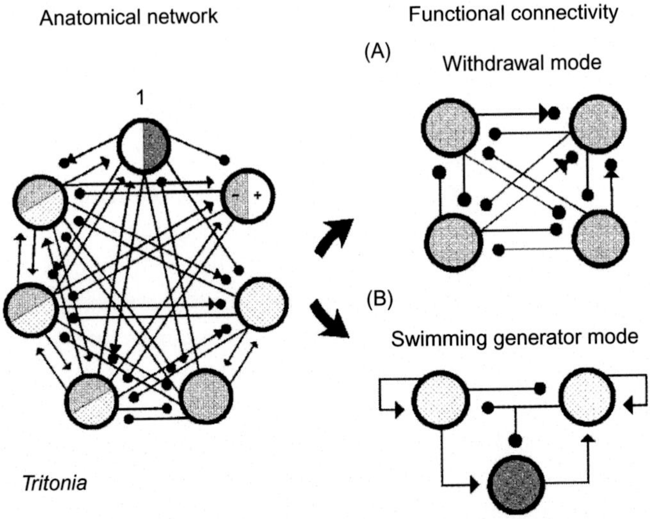

Figure 3.3 Neuronal network of the *Tritonia*, anatomical monosynaptic connectivity (excitatory and inhibitory) and the two possibilities for functional-behavioural networks acting as (A) withdrawal mode and (B) swimming generator mode. The neuron 1 (top, left) activation–deactivation is the first action to produce the functional and, therefore, anatomical reorganization shift, changing the animal behaviour. *Modified from Getting, P.A. 1989. Emerging principles governing the operation of neural networks. Ann. Rev. Neurosci., 12: 185–204.*

anatomical network may be configured into any one of several modes... The term modes is intended to imply a manner in which a network processes information or generates an output pattern'. When afferent or modulating inputs alter the properties of the basic constituents of a set of networks, a transition among modes may occur, for example, in the present case passing from wakefulness to sleep. A neuron, as a basic constituent of a network or cell assembly, may fire an action potential or not or may increase or decrease firing while still belonging to the same network although participating in a new particular function. However, increasing firing does not necessarily mean that a cell is subserving two different processes, such as sleep or waking. In the sleep case for instance, a recorded neuron may belong to a network in which, although engaged in sleep, should play a decreasing firing *mode*.

The *hub* neurons (Bonifazi et al., 2009) have very extensive axonal arborizations projected over larger distances and make greater numbers of and stronger synaptic connections than nonhub neurons. Finally, they are also more responsive to inputs and quicker to fire action potentials themselves, placing them in a position to orchestrate the responses of an entire network. Thus, a hub may quickly change from one network to others, that is, on passing from wakefulness to sleep, introducing to my new concept of a neuronal/network shift that determines the sleep neurophysiologic characteristics (Velluti and Pedemonte, 2010, 2012).

The difference introduced by cell 1 in the *Tritonia* movements by functionally constructing new networks (Fig. 3.3) could be produced, in a complex brain, by a summation of factors. After some unknown signal (s), such as decreasing light intensity, a group of neuronal networks/cell assemblies progressively begin to change the system into a *sleeping mode.* This assumption is partially supported by the observation that when a human or animal is entering sleep, the many variables usually recorded never occur in synchrony, but instead appear with seconds of difference among them (EEG slow activity, electromyogram decrement, eye movements, hippocampal theta rhythm frequency and amplitude, heart rate shifts, blood pressure changes, respiration rate alterations and so on).

Sleep, on the other hand, is not a function but a completely different central nervous system state. Hence, sleep means a whole change of networks/cell assemblies, a new cooperative interaction among them, considering that a single network may subserve different functions when asleep or awake. In addition to the auditory unit firing in guinea pig waking and sleep (Chapter 5, Auditory Unit Activity in Sleep), there is another available demonstration of the neuronal networks subserving auditory changes on passing to sleep, which is shown in Chapter 5, Auditory Unit Activity in Sleep, Fig. 17.5: the auditory evoked magnetoencephalography dipole location shift on the *planum temporale*, indicative of the fact that another set of neurons becomes engaged in the auditory signal processing in human sleep stage II.

New sets of neuronal networks/cell assemblies are reorganized on passing from waking to sleep, therefore constituting a different state, while they continue to be receptive to sensory incoming information.

REFERENCES

Bonifazi, P., Goldin, M., Picardo, M.A., Jorquera, I., Cattani, A., Bianconi, G., et al., 2009. GABAergic *hub* neurons orchestrate synchrony in developing hippocampal networks. Science 326 (5958), 1419–1424.

Carr, C.E., Konishi, M., 1990. A circuit for detection of interaural time differences in the brain stem of the barn owl. J. Neurosci. 10, 3227–3246.

Getting, P.A., 1989. Emerging principles governing the operation of neural networks. Annu. Rev. Neurosci. 12, 185–204.

Liberman, T., Velluti, R.A., Pedemonte, M., 2009. Temporal correlation between auditory neurons and the hippocampal theta rhythm induced by novel stimulations in awake guinea pigs. Brain Res. 1298, 70–77.

Rieke, F., Warland, D., de Ruyter, R., Bialek, W., 1997. Spikes. Exploring the Neural Coding. Massachusetts Institute of Technology, Massachusetts, CA.

Sakurai, Y., 1999. How do cell assemblies encode information in the brain? Neurosci. Biobehav. Rev. 23, 785–796.

Segundo, J.P., Moore, G.P., Stensaas, L.J., Bullock, T.H., 1963. Sensitivity of neurons in *Aplysia* to temporal pattern of arriving impulses. J. Exp. Biol. 40, 643–667.

Shannon, C.E., 1948. A mathematical theory of communication. Bell Sys. Tech. J. 27, 379–423.

Velluti, R.A., Pedemonte, M., 2010. Auditory neuronal networks in sleep and wakefulness. Int. J. Bifurcat. Chaos 20, 403–407.

Velluti, R.A., Pedemonte, M., 2012. Sensory neurophysiologic functions participating in active sleep processes. Sleep Sci. 5 (4), 103–106.

CHAPTER 4

Auditory Information Processing During Sleep

Incoming sensory information is a relevant factor to be considered when waking or sleep are under question. The following experimental data introduce the subject leading to further analysis of the relationships between audition and sleep.

DEAF PATIENTS AND SPECIAL SLEEP ANALYSIS

To properly demonstrate the effect of auditory input during sleep of intracochlear implanted patients, the following approach was developed. Four implanted deaf patients were recorded during four nights: two nights with the implant off, with no auditory input, and two nights with the implant on, that is, with normal auditory input, being only the common night sounds present, with no additional auditory stimuli delivered. The sleep patterns of another five deaf people were used as controls, exhibiting normal sleep organization. Moreover, the four experimental patients with intracochlear devices and the implant off also showed normal sleep patterns. On comparison of the night recordings with the implant on and off, a new sleep organization was observed for the recordings with the implant on, suggesting that brain plasticity may produce changes in the sleep stage percentages while maintaining the ultradian rhythm (Fig. 4.1).

During sleep with the implant on, the analysis of the electroencephalographic delta, theta and alpha bands in the frequency domain, using the fast Fourier transform, revealed a diversity of changes in the power originated in the contralateral cortical temporal region.

Several lines of evidence support the notion that the auditory activity encompasses a special relationship with sleep. Incoming auditory signals may introduce changes in sleep characteristics, and in turn, sleep may introduce changes in every auditory nucleus via the auditory efferent system or at a cortical level. For instance, the auditory is a telereceptor

DOI: https://doi.org/10.1016/B978-0-12-810476-7.00004-X

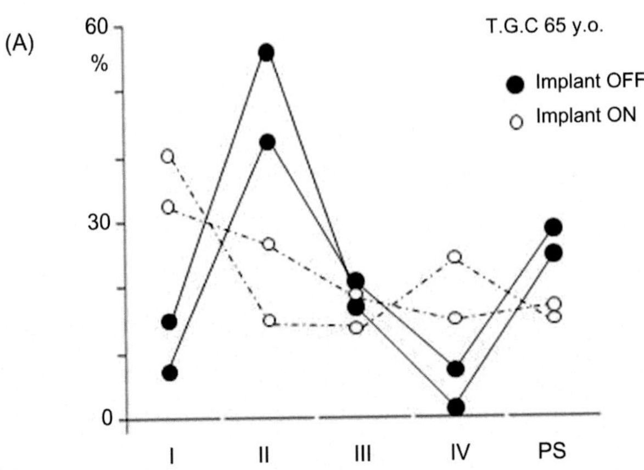

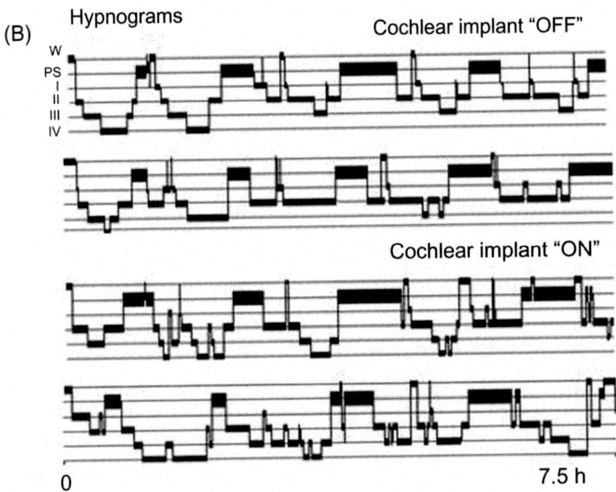

Figure 4.1 Example of sleep stages organization in a deaf implanted patient. (A) The filled circles show control recording nights with the implant off, that is, recorded as a deaf patient, and the open circles, implant on, that is, as a hearing patient. The decrease of stage II and paradoxical sleep (PS), together with the stage IV percentage increase, are seen once more. (B) The hypnograms display the whole night sleep ultradian cycle recordings (7.5 h), two nights with the implant off and two nights with the implant on. No shifts in the sleep ultradian cycles were observed. Sleep stages are I, II, III and IV (equivalent to slow wave sleep [SWS] or non-REM sleep) and PS (equivalent to REM sleep). *Modified from Velluti, R.A., Pedemonte, M., Suárez, H., Bentancor, C., Rodriguez-Servetti, Z., 2010. Auditory input modulates sleep: an intra-cochlear-implanted human model. J. Sleep. Res. 19, 585–590.*

system continuously active that informs the central nervous system (CNS) about the environment during wakefulness and sleep (Velluti et al., 2010).

WITHOUT SENSORY INPUT

The surgical section of the olfactory, optic, statoacoustic and trigeminal nerves; one vagus nerve; and the spinal cord posterior paths in cats, a quasitotal deafferentation, carried out by Vital-Durand and Michel (1971), reconfirmed a decrement in motor activity previously reported by Galkine (1933) and Hagamen (1959) with a predominant sphinx head and body position. Thus, the decreased contact with the external world may induce particular behaviours and may also result in the lack of behavioural initiatives. The former data once cited led to the erroneous idea of a 'continuous sleep', 99% of the time according to Hagamen (1959), in an almost totally sensory-deprived cat. When studied with polygraphic controls (Vital-Durand and Michel, 1971), the animals under quasitotal deafferentation condition revealed a sleep-waking cycle showing different characteristics:

1. The waking time was reduced from 44.9% to 18.5% and when asleep the cats could be awakened easily at any moment. The electroencephalogram (EEG) characteristics were normal with the exception of slow wave bursts in the visual cortex (Kasamatsu et al., 1967).
2. The time spent in slow wave sleep (SWS) was reduced from 41.7% to 29.6%. An almost constant 'somnolence' was described associated with the sphinx position and a sequential fast and slow EEG activity. On the other hand, the hippocampus and amygdala activity was that of a quiet wakefulness (W) indicative of a distinct state, both from a behavioural and from a bioelectrical viewpoint.
3. The total amount of paradoxical sleep (PS) was slightly diminished (from 13.4% to 11.2%) with normal episodes length and frequency. PS showed the characteristic phasic signs as eye movements, middle-ear muscles and muscular twitches, present prior to the cervical spinal cord lesion. Muscle activity inhibition as well as cortical EEG activation, hippocampal theta rhythm and ponto-geniculo-occipital waves in the visual cortex were also present. Moreover, the sensorimotor cortex had an active, desynchronized EEG and the hippocampus exhibited a normal theta rhythm; the visual cortex showed short periods of flat EEG between bursts of high amplitude fast activity.

After sensory deprivation in cats, a new behavioural state seems to develop. Considering the behaviour and the bioelectrical activities recorded, a new state could be described under what was called *somnolence*. The cat sphinx behaviour observed was associated with a peculiar occipital EEG and normal subcortical activity, thus characterizing what I am postulating as a new although abnormal brain state dependent on the lack of sensory input.

Total auditory deprivation in guinea pigs by surgical removal of both cochleae enhances SWS and PS in similar proportion, while reducing wakefulness (Pedemonte et al., 1996). The SWS and PS increments cited previously were determined mainly by an increase in the number of episodes, but there was no change in the duration of a single episode. The authors assert that the relative isolation from the outside world may be in part cause of the change observed in deaf guinea pigs. Therefore, the elimination of an input to a complex set of networks, such as the ones that regulate the sleep-waking cycle, would introduce functional shifts particularly if such input has some significance, as appears to happen in the case of the behaviour under study: wakefulness and sleep.

Human sensory deprivation experiments are different, since they should be better considered as a reduction of sensory input as much as possible, which leads to the notion that, when a human subject is placed in an environment without patterned and changing stimulation, they may fall into a state of lowered arousal and sleep (Zubek, 1969).

EVOKED ACTIVITY

Two types of evoked responses can be recorded in humans: near-field and far-field evoked potentials. In animal experimentation a local-field activity can be recorded with electrodes placed into the nuclei or cortex. The evoked potential recorded from the scalp represents a mixture of both near-field and far-field potentials. The evoked potentials provide information about the sensory processing, to localize a lesion site, or to add data about the maturation or of an aged brain. Moreover, two types of response are usually distinguished in humans, *sensory* and *cognitive* event-related potential (ERP).

During sleep, a normal reaction to any suprathreshold sensory stimulation is a return to an awake condition. Moreover, on sensory stimulation, a special evoked EEG pattern was described in SWS (Loomis et al., 1938) designated as the K-complex. Apart from appearing spontaneously,

K-complexes were observed in response to sensory stimulation such as visual, somesthetic, or auditory stimuli, the last being the most effective (Halász and Ujszászi, 1988; Bastien et al., 2002). The K-complex response to auditory stimulation was large and less variable during stage II sleep, while during SWS (III–IV) there were no sensory-evoked modifications of the electrical activity (Davis et al., 1939). Spontaneous K-complex may be due to interoceptive stimuli and some components of the evoked K-complexes show habituation in response to repetitive stimulation (Bastuji and García-Larrea, 2005).

SENSORY-EVOKED RESPONSES

Sensory data input and data integration is not abolished during sleep. Several electrophysiological reports from human and animal experimentation as well as the easy awakening with significant sounds and the incorporation of stimuli into dreams are results indicative of the conservation of some kind of processing in sleep, surely different from the one observed during W.

Human auditory responses recorded from the vertex have been reported by several investigators using similar approaches and obtaining similar results (Fig. 4.2). In all subjects the auditory evoked responses exhibit major changes on passing from the wake state to the stages I, II and SWS and a consistent increase in peak-to-peak amplitude was present

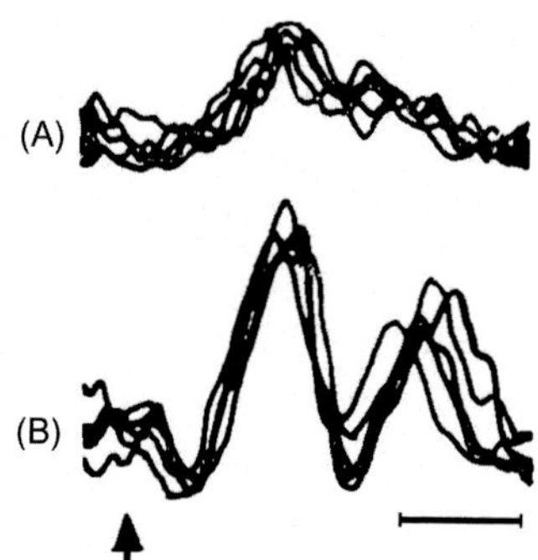

Figure 4.2 Auditory evoked potentials during wakefulness (A) and sleep (B) in humans. Superposition of five averaged responses showing amplitude increment and wave complexity on passing into sleep. A mixture of both near-field and far-field potentials are part of this response. Click stimuli at 1/s. *Modified from Vanzulli, A., Bogacz, J., García-Austt, E., 1961. Evoked responses in man. III. Auditory response. Acta Neurol. Latinoamer. 7, 303–308.*

while, during PS, the amplitude was decreased, approximately to that of the wake state. The latter waves of the response were of longer latency during both sleep phases, SWS and PS (Vanzulli et al., 1961; Williams et al., 1963; Davis and Yoshie, 1963; Weitzman and Kremen, 1965; Ornitz et al., 1967).

HUMAN AUDITORY FAR-FIELD EVOKED POTENTIALS

In a comprehensive review of the evoked potentials and information processing during sleep, some interesting conclusions were drawn (Campbell et al., 1992). The experimental data, gathered using the far-field potential recording technique in humans, showed no sleep effects on the brainstem auditory evoked potentials (Amadeo and Shagass, 1973; Picton et al., 1974; Osterhammel et al., 1985; Bastuji et al., 1988). However, Bastuji et al. (1988) reported small brainstem auditory evoked potential (BAEP) latency changes during nocturnal sleep. In addition, the constancy of the response was maintained whether sound stimuli were either of high or low intensity (Campbell and Bartoli, 1986).

The BAEP are far-field potentials technically recorded distant from the brainstem nuclei, being a coarse representation of auditory function that do not reflect the effects of sleep. On the other hand, the auditory nerve compound action potential (cAP) in sleep, described in guinea pigs (Velluti et al., 1989), or the effects of sleep on the guinea pig auditory nuclei unitary firing have shown significant cAP amplitude and unit firing shifts (see the review by Velluti, 1997; Pedemonte and Velluti, 2005a). The two technical approaches, human and animal, have led to different results that deserve more research for the reason that using a more sensitive technique, sleep-related changes will also appear in humans.

Kräuchi et al. (2006) reported that proximal as well as distal increases in skin temperature are present during naptime. These reports are indicative of a great difference in temperature control during both night and nap sleep. Then, we began to study the BAEP during nap-sleep-like in human subjects. Volunteers with no pathology were stimulated with alternating clicks (80 dB SPL, 10/s) and BAEPs were recorded together with online sleep polysomnographic control, under chloral hydrate at a hypnotic dose. The experimental design was to record during the afternoon nap (from 1:00 to 4:00 p.m.) with monitored skin temperature and evoked responses studied; we found a significant increase in wave V latency not related to body temperature but to sleep N2 (Fig. 4.3). Waves

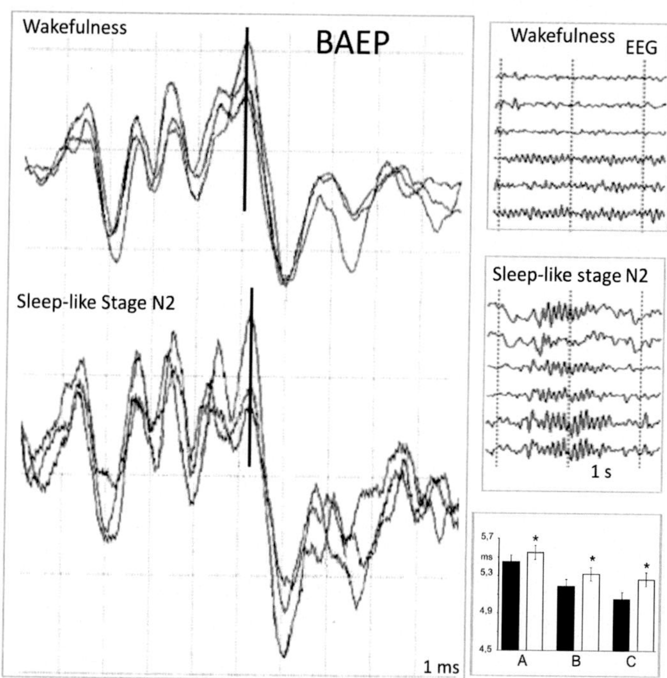

Figure 4.3 Brain auditory evoked potentials (BAEPs), superpositions of three subjects (A, B and C). Right side EEG showing a wakefulness recording (upper), while in the middle one a sleeplike N2 stage, under chloral hydrate, is depicted. The lower right corner shows the corresponding latencies change significance ($P < .05$; Student *t* test). Black bars, wakefulness; open bars, sleeplike N2. *Modified from Medina-Ferret et al., 2009, Pedemonte, M., Medina-Ferret, E., Velluti, R.A., 2016. Sensory processing in sleep: an approach from animal to human data. In: P. Perumal (Ed.), Synopsis of Sleep Medicine, APP, CRS (Chapter 22), pp. 379–396 (Pedemonte et al., 2016).*

I–V (10 ms), also studied in the frequency domain (FFT, showed significant decreases in the power spectra of 400, 500 and 600 Hz during sleep-like stage N2.

Another phenomenon also points out sleep actions on the auditory receptor itself, namely, the transiently evoked oto-acoustic emissions (sounds emitted by the cochlea reflecting the cochlear outer hair cell motility controlled by the efferent system, i.e., the CNS). It has been reported in humans as being modified by sleep in general although independent of the sleep phase (Froehlich et al., 1993).

Data regarding the sleep effects on middle latency auditory evoked potentials (potentials perhaps arising from the reticular formation, thalamus and primary cortex) are not consistent. While early studies indicated that these components were either not affected or only slightly affected by sleep, more recent reports showed marked changes most notably on the later evoked potential components (Mendel and Goldstein, 1971; Osterhammel et al., 1985; Erwin and Buchwald, 1986; Ujszászi and Halász, 1986), although Campbell et al. (1992) suggested that these waves were attenuated only with fast stimulation.

On the other hand, auditory steady-state response in infant testing may provide additional audiometric information for accurately predicting the hearing sensitivity, and this is essential for the management of infants with severe to profound hearing loss (Lee et al., 2008).

LONG LATENCY EVOKED POTENTIALS

The later components of the evoked potential, also called the *slow potentials* or *late auditory evoked responses*, are altered most during sleep.

Long latency W auditory evoked potentials are biphasic, negative-positive complexes, N1–P2, at 100–150 ms poststimulus. The neural generators of the N1–P2 responses are not well known. However, both modelling and intracranial recordings have located the sources in the secondary auditory cortical areas.

Sleep is characterized by a latency delay and amplitude decrease of the N1, associated with an enhancement of P2 (see Campbell et al., 1992). Shifts in N1–P2 during sleep onset are fast and time locked to sleep entrance, which has been proposed as a marker of sleep onset (Ogilvie et al., 1991; Campbell et al., 1992). After sleep entrance, the N1 and P2 modifications persist with little change during both SWS and PS (see Bastuji and García-Larrea, 1999, 2005).

COGNITIVE EVENT-RELATED POTENTIALS

ERPs can be recorded by introducing deviant, unexpected stimuli within a stimulus train. The ERPs may be part of cognitive processes: capacity of discrimination, attention, memory and so on. High-level sensory integration, using equal stimulating paradigms, has validated the possibility of such processing continuity during the full sleep-waking cycle.

P300

The P300, a human bioelectrical component, appears in association with deviant stimuli delivered at random in a stimulating regular train, that is, the discrimination of an odd-ball stimulus. P300, a positive component with the highest amplitude between 220 and 350 ms, is usually associated with attention and discrimination. The possible P300 generators are the prefrontal cortex, the temporal and parietal cortices, the hippocampus and the cingulate gyrus, all contributing to its generation (Escera et al., 2000; Baudena et al., 1995; Brázdil et al., 1999, 2001). The P300 latency increases and amplitude decreases from wakefulness through sleep (Wesensten and Badia, 1988; Nielsen-Bohlman et al., 1991; Harsh et al., 1994; Atienza et al., 2001; Cote, 2002). Apparently, P300 can be recorded during the transition from wakefulness to stage I sleep, reappearing during PS. The lack of the frontal component of P300 may support the notion of semiautomatic, nonconscious detection of stimuli during sleep.

The averaged ERPs (P300) to regular and deviant tones during W, stage I and PS are shown in Fig. 4.4 exhibiting shifts on passing into

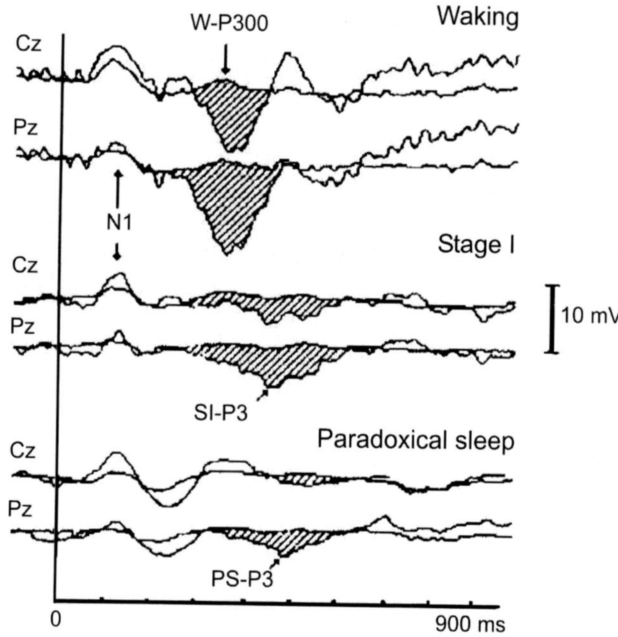

Figure 4.4 Event-related potentials (ERPs), P300, obtained with pure tones in a classical odd-ball paradigm. P300 in one subject during waking and sleep stage I and PS. *Modified from Bastuji, H., García-Larrea, L., Bertrand, O., Mauguiere, F., 1988. BAEP latency changes during nocturnal sleep are not correlated with sleep stages but with body temperature variations. Electroenceph. Clin. Neurophysiol. 70, 9–15.*

sleep. Moreover, although with some differences (not shown), changes are also present in stage II and SWS.

The results exhibited are indicative of a capability of the sleeping brain to detect equally physical stimuli among a group of regular ones, thus reacting to novelties. These facts do not necessarily imply that their underlying processes are equivalent to those of occurring in wakefulness.

Mismatch Negativity

Mismatch negativity (MMN) is elicited by deviant stimuli occurring at random within a stream of regular tones. It is acknowledged as reflecting an automatic detection of a series of new sensory input that does not 'match' the neuronal representation of the regular stimuli, an automatic comparison process (Näätänen, 1995). MMN is considered to reflect nonconscious activity of the sensory processing, related to the temporal and frontal networks activation (Fig. 4.5). The analysis of this response in sleep is a step forward in studying sensory processing. While the persistence of a system of deviance detection during stages I, II and SWS is supported by a bulk of convergent studies, the available evidence suggests that such detection does not involve the generation of a cerebral response comparable to the waking MMN (Atienza et al., 2002).

Contrary to results in stages I, II and SWS (Campbell et al., 1992), other investigators reported a genuine MMN when recording the stage II and during PS (Loewy et al., 1996; Atienza et al., 1997). The PS MMN appeared to be of similar latency but smaller in amplitude relative to its waking counterpart. Interestingly, this component has been reported

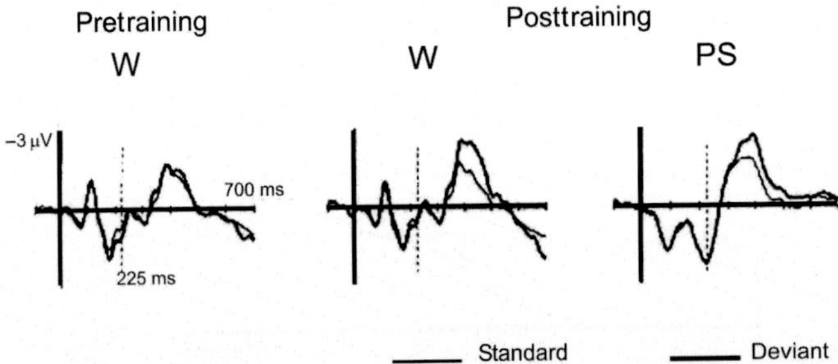

Figure 4.5 Mismatch negativity. Grand average waveforms to standard (thick line) and deviant (thin line) stimuli during wakefulness (W) pretraining and after training, and posttraining during PS. *Modified from Atienza, M., Cantero, J.L., Escera, C., 2001. Auditory information processing during human sleep as revealed by event-related brain potentials. Clin. Neurophysiol. 112, 2031–2045.*

during quiet sleep (equivalent to adult SWS) in newborns (Cheour et al., 2002).

However, a possible influence of technical factors hampering the recordings, such as an unfavourable signal-to-noise ratio relative to waking, should not be dismissed, since this very small component may be lost within the much higher sleep negativities. However, during sleep the auditory cortex can maintain the ability to organize the arrived information although with great differences from the corresponding one in wakefulness. The MMN system activates memory when it does not match related immediate information (Atienza et al., 2002). Moreover, the well-known bottom-up influences (Velluti and Pedemonte, 2002) are added to the sleep top-down actions, shown particularly in the auditory system through its efferent fibres system (Velluti, 1997).

EXPERIMENTAL ANIMAL STUDIES OF EVOKED POTENTIALS

The amplitude of the averaged auditory nerve cAP and the microphonic potentials, in response to clicks and tone bursts, has been reported to change throughout the guinea pig sleep-waking cycle (Velluti et al., 1989). Both potentials, recorded in response to low intensity sound stimulation during sleep phases, changed in a roughly parallel fashion, increasing amplitude during SWS and decreasing during PS to amplitudes similar to those observed in W (Fig. 4.6). These results were more recently repeated using new devices and applying new techniques while still using guinea pigs (Pedemonte et al., 2004).

This modulation of the auditory peripheral potentials by the sleeping brain (through activation of the efferent terminals on the micromechanical properties of the coupled efferent bundle hair cells and direct actions on the afferent fibres) resulted in, among other things, a modulation/gating of the auditory input ensuing in a facilitation during SWS, a reduction of the input during PS and variable amplitudes during W. A corollary to this statement is that the changes observed in the peripherally recorded auditory potentials during W and sleep phases may condition, at least in part, the described shifts in the responses evoked upstream in the pathway, as far as the auditory cortex itself.

The CNS auditory evoked potentials changes associated with sleep, wakefulness or anaesthesia have been reported in cats by several authors (Huttenlocher, 1960; Jouvet, 1962; Pradham and Galambos, 1963; Teas and Kiang, 1964; Herz, 1965; Herz et al., 1967; Wickelgren, 1968;

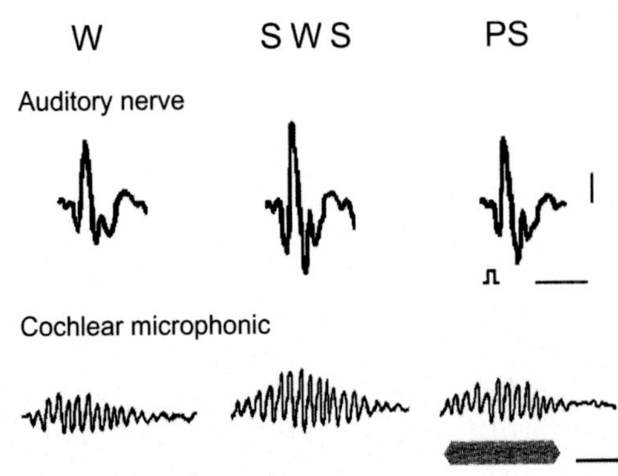

Figure 4.6 Auditory nerve compound action potential (cAP) and cochlear microphonic (CM) during wakefulness (W), SWS and PS in the guinea pig. Both signals were recorded from the round window with an implanted macroelectrode. The CM was evoked by a pure tone burst (1 kHz) and the cAP by a click (0.15 ms). CM and cAP were averaged ($n = 30$) and in both cases the amplitude increased during SWS and decreased during PS in comparison with a quiet waking period. Calibration bars are cAP, 5 ms, 50 μV; CM, 5 ms, 200 μV. *Modified from Velluti, R.A., Pedemonte, M., García-Austt, E., 1989. Correlative changes of auditory nerve and microphonic potentials throughout sleep. Hear. Res. 39, 203–208.*

Petrek et al., 1968) and in rats (Hall and Borbély, 1970). Several discrepancies among the reports by different authors were observed in cats as well as in humans. For example, some investigators studying cats reported a decrease of the late surface negative wave during SWS (Huttenlocher, 1960; Teas and Kiang, 1964) while Herz (1965) and Herz et al. (1967) reported an increase in amplitude of such waves; this latter result was corroborated by Wickelgren (1968).

Continuous measurements of evoked responses in long, well-controlled experiments were made by Hall and Borbély (1970) at different auditory cortical depths, in the medial geniculate body, the thalamic reticular formation and the hippocampus (Fig. 4.7). The cortical results showed that the evoked 'potentials recorded from a depth of 1–1.5 mm, all components of the average waveform being larger during SWS than in waking and low-voltage fast sleep'. Responses from the cortical surface are reported, including increases of the first and second positive waves as well as of the late negative wave, during SWS. The early components of both the surface and deep responses were similar in W and in PS.

The medial geniculate evoked response in sleep was represented only by an amplitude reduction during PS. Neurons of the medial geniculate

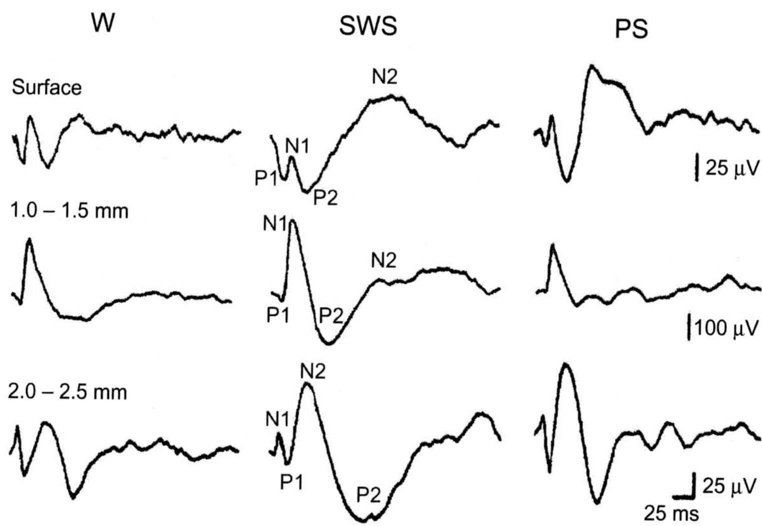

Figure 4.7 Click-evoked averaged cortical responses in rats, recorded from surface and from depths of 1.0–1.5 and 2.0–2.5 mm in the auditory cortex during wakefulness (W), SWS and PS. Positive changes of potentials are indicated by downward deflections. *Modified from Hall, R.D., Borbély, A.A., 1970. Acoustically evoked potentials in the rat during sleep and waking. Exp. Brain. Res. 11, 93–110.*

nucleus showed evoked firing shifts mainly decreasing on passing from W to SWS, while the spatial receptive field was preserved indicating that the information sent to cortical cells may carry significant content (Edeline et al., 2000, 2001).

The hippocampal and the thalamic reticular formation recordings showed increased late positive waves during SWS that switched to a smaller amplitude during both W and PS.

Thus, according to Hall and Borbély (1970), the late negative components are larger in sleep than during wakefulness in the rat, which agrees with the findings of the Herz (1965, 1967) and also with data from Wickelgren (1968), in cats.

It has been shown (Baust et al., 1964) that middle-ear muscles contraction during PS, often accompanying rapid eye movements as phasic activity, reduced the auditory input. Although many authors did not take this possibility into account, all recording in our guinea pigs experiments were carried out without middle-ear ossicles. Besides, it was suggested that recording in cats with tenotomized middle-ear muscles, the middle-ear mechanisms are not responsible for reduced medial geniculate evoked responses during PS (Berlucchi et al., 1967; Wickelgren, 1968). The hippocampal evoked potentials reported in rats (Hall and Borbély, 1970) were quite similar to those evoked in cats; in both species the late positive wave increased markedly during SWS.

HUMAN MAGNETOENCEPHALOGRAPHIC EVOKED ACTIVITY RECORDINGS IN SLEEP

With a high temporal resolution, on the order of milliseconds, EEG and the MEG are the techniques with the capability to show changes of the early cortical data processing. The magnetic fields recorded from the human brain on the scalp are very small, and special devices are needed to detect such signals. Furthermore, MEG has the great advantage of detecting within the millimetre range the source localization of a response; that is, it is possible to show the dipole position over the cortex in a very precise way. This is one important finding when analysing the responses to sound during wakefulness and sleep (Kakigi et al., 2003).

The four components of the human auditory evoked magnetic field are exhibited in Fig. 4.8. The M150 and M200 waves show a clear increment in amplitude on passing from W to sleep stages I and II.

Regarding the dipole location, differences appear during sleep stages I and II: the M50 dipole wave exhibits a more anterior and lateral position on the auditory cortex. The M100 wave dipole also presents a similar change in location (Kakigi et al., 2003). The dipole cortical position shifts constitute an important change in relation to the auditory processing as well as to the sleep organization in particular, as I discuss this later, in Chapter 5, Auditory Unit Activity in Sleep.

HUMAN AUDITORY BRAIN AREAS IMAGING

Sensory processing in sleep exhibited a new look after the introduction of the imaging studies, particularly the functional magnetic resonance technique (fMRI), although not all the approaches are unanimous in their results (Braun et al., 1997; Born et al., 2002; Portas et al., 2000; Czisch et al., 2002; Tanaka et al., 2003). The fMRI analysis by Portas et al. (2000) exhibited the auditory stimuli in SWS producing a significant bilateral activation in the auditory cortices, thalamus and caudate in comparison to wakefulness (Fig. 4.9). On the other hand, Czisch et al. (2002, 2004), also using fMRI, reported a reduced activation (not a lack of it) in

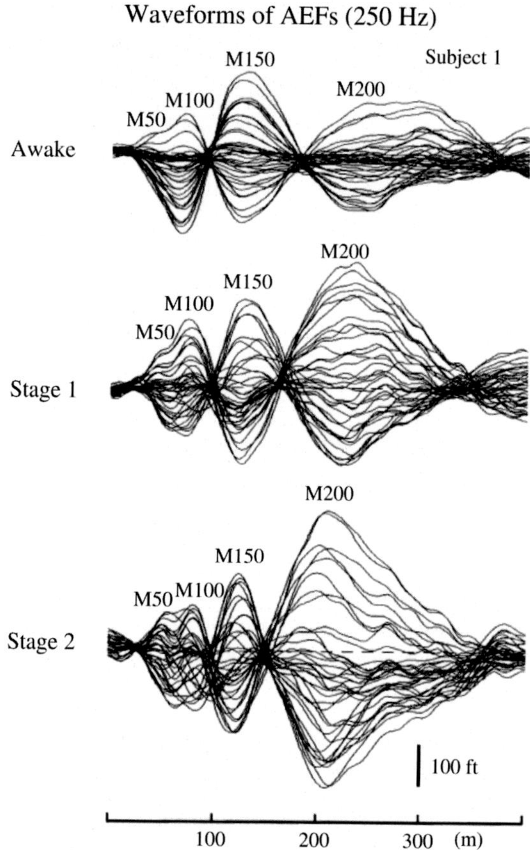

Figure 4.8 Human auditory evoked MEG response during wakefulness and sleep stages I and II. A pure tone of 250 Hz was delivered to the right ear and the magnetometer was placed on the left hemisphere (position C3). Thirty-seven superimposed waveforms at each stage are shown. Four main components (M50, M100, M150 and M200) were identified in each stage. Although M50 and M100 showed no definite change during the two stages of sleep, M150 and M200 were significantly enhanced in both sleep stages. However, M50, M100 and M200 were significantly prolonged to 1000 or 4000 Hz. *Modified from Kakigi, R. Naka, D., Okusa, T., Wang, X., Inui, K., Qiu, Y., et al., 2003. Sensory perception during sleep in humans: a magnetoencephalograhic study. Sleep. Med. 4, 493–507.*

the auditory cortex during sleep stages I, II and SWS. Some of the research groups cited concluded that the decreased response may protect the sleeping brain from the arousing effects of external stimulation during sleep. These results are not congruent at all with the notion that ~50%

Tone-burst stimulation during sleep

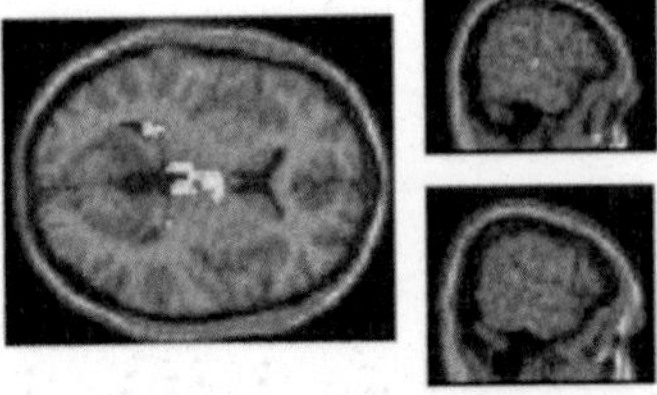

Name stimulation during sleep

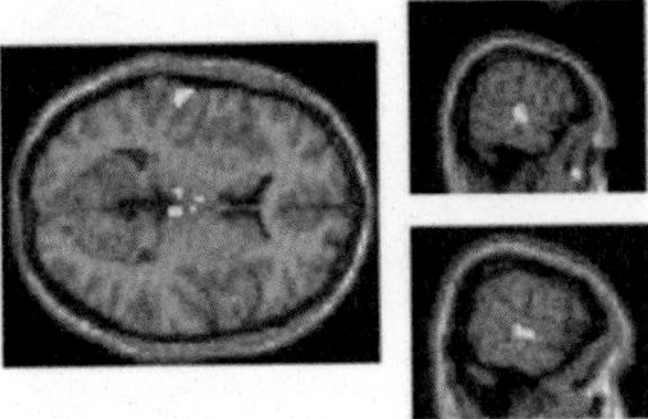

Figure 4.9 Brain areas activation both during auditory stimulation with a nonsignificant sound, tone bursts, and with a significant sound, the person's name, during sleep and in comparison to wakefulness. *Modified from Portas, C.M., 2005. Cognitive aspects of sleep. In: Parmeggiani, P.L., Velluti, R.A. (Eds.), The Physiologic Nature of Sleep. Imperial College Press, London, pp. 535–569.*

of auditory cortex neuron in guinea pigs continue firing as during wakefulness on passing to SWS, as reported by Peña et al. (1999).

In 1830 the German physiologist Burdach (cited by Portas, 2005) wrote, 'an indifferent word does not arouse the sleeper, but if called by name he awakens ... the mother awakens to the faintest sound from her child ... the miller wakes when the mill stops ... hence the psyche differentiates sensations during sleep'. The first researchers to demonstrate, in a controlled experiment, that personal names evoke more K-complexes in the sleeper than any other name or sounds of same intensity were Oswald et al. (1960). Moreover, the analysis of evoked potential showed a differential cognitive response to the presentation of the subject's own name (a significant sound) during sleep stage II and PS (Berland and Pratt, 1995; Perrin et al., 1999).

The fMRI analysis of data associated with EEG activity exhibited a specific response to the subject's own name during SWS (Fig. 4.9). In addition, presentation of the subject's own name during sleep was associated with selective activation of the left amygdala and left prefrontal cortex (Portas, 2005; Portas et al., 2000).

In humans, the cortex activity changes are not homogeneous. When compared to wakefulness, the less active areas are located in associative cortices of the frontal and parietal cortices, while the primary cortex is the least or not deactivated (Braun et al., 1997; Maquet et al., 1997; Andersson et al., 1998; Kajimura et al., 1999).

The sleeping brain is a totally different brain that involves activated–deactivated areas to complete its many physiological aspects. Sleep is a state still responsive to external stimuli (Velluti and Pedemonte, 2002; Edeline, 2003; Pedemonte and Velluti, 2005b). ERPs and also the evoked potentials (far and local-field potentials) recorded in humans and animals have demonstrated that environmental as well as internal, body, stimuli are processed in sleep.

The many results obtainable suggest that the processing of external stimuli can go beyond the primary cortices during stages I, II and SWS. The neurophysiological actions by which salient stimuli can recruit associative cerebral areas during sleep remain unclear (Maquet et al., 2005).

In comparison with wakefulness, differential patterns of regional cerebral blood flow activity are observed during stages I, II, SWS and PS in humans. Maquet et al. (2005) suggest, 'the neural populations recently challenged by a new experience are reactivated and increase their functional connectivity during the post-training sleep episodes, suggesting the off-line processing of recent memory traces in sleep'.

REFERENCES

Amadeo, M., Shagass, C., 1973. Brief latency click-evoked potentials during waking and sleep in man. Psychophysiology 10, 244–250.

Andersson, J.L., Onoe, H., Hetta, J., Lidstrom, K., Valind, S., Lilja, A., et al., 1998. Brain networks affected by synchronized sleep visualized by positron emission tomography. J. Cereb. Blood. Flow. Metab. 18, 701–715.

Atienza, M., Cantero, J.L., Gomez, C.M., 1997. The mismatch negativity component reveals the sensory memory during REM sleep in humans. Neurosci. Lett. 237, 21–24.

Atienza, M., Cantero, J.L., Escera, C., 2001. Auditory information processing during human sleep as revealed by event-related brain potentials. Clin. Neurophysiol. 112, 2031–2045.

Atienza, M., Cantero, J.L., Domínguez-Marín, E., 2002. Mismatch negativity (MMN): an objective measure of sensory memory and long-lasting memories during sleep. I. J. Psychophysiol. 46, 215–225.

Bastien, C., Croewley, K.E., Colrain, I.M., 2002. Evoked potential components unique to non-REM sleep: relationship to evoked K-complexes and vertex sharp waves. Int. J. Psychophysiol. 46, 257–274.

Bastuji, H., García-Larrea, L., 1999. Evoked potentials as a tool for the investigation of human sleep. Sleep Medicine Rev. 3, 23–45.

Bastuji, H., García-Larrea, L., 2005. Human auditory information processing during sleep. In: Parmeggiani, P.L., Velluti, R.A. (Eds.), The Physiologic Nature of Sleep. Imperial College Press, London, pp. 509–534.

Bastuji, H., García-Larrea, L., Bertrand, O., Mauguiere, F., 1988. BAEP latency changes during nocturnal sleep are not correlated with sleep stages but with body temperature variations. Electroenceph. Clin. Neurophysiol. 70, 9–15.

Baudena, P., Halgren, E., Heit, G., Clarke, J.M., 1995. Intracerebral potentials to rare target and distractor auditory and visual stimuli. III. Frontal cortex. Electroenceph. Clin. Neurophysiol. 94, 251–264.

Baust, W., Berlucchi, G., Moruzzi, G., 1964. Changes in the auditory input in wakefulness and during the synchronized and desynchronized stages of sleep. Arch. Ital. Biol. 102, 657–674.

Berland, I., Pratt, H., 1995. P300 in response to the subject's own name. Electroenceph. Clin. Neurophysiol. 96, 472–474.

Berlucchi, G., Munson, J.B., Rizzolatti, G., 1967. Changes in click evoked responses in the auditory system and the cerebellum of free-moving cats during sleep and waking. Arch. Ital. Biol. 105, 118–135.

Born, A.P., Law, I., Lund, T.E., Rostrup, E., Hanson, L.G., Wildschiodtz, G., et al., 2002. Cortical deactivation induced by visual stimulation in human slow-wave sleep. Neuroimage 17, 1325–1335.

Braun, A.R., Balkin, T.J., Wesenten, N.J., Carson, R.E., Varga, M., Baldwin, P., et al., 1997. Regional cerebral blood flow throughout the sleep-wake cycle. An H2(15)O PET study. Brain 120, 1173–1197.

Brázdil, M., Rektor, I., Dufek, M., Daniel, P., Jurák, P., Kuba, R., 1999. The role of frontal and temporal lobes in visual discrimination task – depth ERP studies. Neurophysiol. Clin. 29, 339–350.

Brázdil, M., Rektor, I., Daniel, P., Dufek, M., Jurak, P., 2001. Intracerebral event-related potentials to subthreshold target stimuli. Clin. Neurophysiol. 112, 650–661.

Campbell, K., Bartoli, E., 1986. Human auditory evoked potentials during natural sleep. Electroenceph. Clin. Neurophysiol. 65, 142–149.

Campbell, K., Bell, I., Bastien, C., 1992. Evoked potential measures of information processing during natural sleep. In: Broughton, R.J., Ogilvie, R.D. (Eds.), Sleep, Arousal, and Performance. Birkhauser, Boston-Basel-Berlin, pp. 89–116.

Cheour, M., Martynova, O., Näätänen, R., 2002. Speech sounds learned by sleeping newborns. Nature 415, 599–600.

Cote, K.A., 2002. Probing awareness during sleep with the auditory odd-ball paradigm. Psychophysiology 46, 227–241.

Czisch, M., Wetter, T.C., Kaufmann, C., Pollmacher, T., Holsboer, F., Auer, D.P., 2002. Altered processing of acoustic stimuli during sleep: reduced auditory activation and visual deactivation detected by a combined fMRI/EEG study. Neuroimage 16, 251–258.

Davis, H., Yoshie, N., 1963. Human evoked cortical responses to auditory stimuli. Physiologist 6, 164.

Davis, H., Davis, P.A., Loomis, A.L., Harvey, E.N., Hobart, G., 1939. Electrical reactions of the human brain to auditory stimulation during sleep. J. Neurophysiol. 2, 500–514.

Edeline, J.-M., 2003. The thalamo-cortical auditory receptive fields: regulation by the sates of vigilance, learning and neuromodulatory systems. Exp. Brain. Res. 153, 554–572.

Edeline, J.-M., Manunta, Y., Hennevin, E., 2000. Auditory thalamus neurons during sleep: changes in frequency selectivity, threshold, and receptive field size. J. Neurophysiol. 84, 934–952.

Edeline, J.-M., Dutrieux, G., Manunta, Y., Hennevin, E., 2001. Diversity of receptive field changes in auditory cortex during natural sleep. Eur. J. Neurosci. 14, 1865–1880.

Erwin, R., Buchwald, J., 1986. Midlatency auditory evoked responses: differential effects of sleep in the human. Electroenceph. Clin. Neurophysiol. 65, 383–392.

Escera, C., Alho, K., Schroger, E., Winkler, I., 2000. Involuntary attention and distractibility as evaluated with event-related brain potentials. Audiol. Neurootol. 5, 151–166.

Froehlich, P., Collet, L., Valaxt, J.L., Morgon, A., 1993. Sleep and active cochlear micromechanical properties in human subjects. Hear. Res. 66, 1–7.

Galkine, V.S., 1933. On the importance of the receptors for the working of the higher divisions of the nervous system. Arkh. Biol. Nauk. 33, 27–55.

Hagamen, W.D., 1959. Responses of cats to tactile and noxious stimuli. Arch. Neurol. Psychiat., Chicago 1, 203–215.

Halász, P., Ujszászi, J., 1988. A study of K-complexes in humans: are they related to information processing during sleep? In: Koella, W.P., Obál, F., Shulz, H., Visser, P. (Eds.), Sleep '86. Gustav Fisher Verlag, Stuttgart-New York, pp. 79–83.

Hall, R.D., Borbély, A.A., 1970. Acoustically evoked potentials in the rat during sleep and waking. Exp. Brain. Res. 11, 93–110.

Harsh, J., Voss, U., Hull, J., Schrepfer, S., Badia, P., 1994. ERP and behavioral changes during the wake/sleep transition. Psychophysiology. 31, 244–252.

Herz, A., 1965. Cortical and subcortical auditory evoked potentials during wakefulness and sleep in cat. In: Akert, K., Bally, C., Shade, J.P. (Eds.), Sleep Mechanisms. Progress in Brain Research. Elsevier, Amsterdam, pp. 63–69.

Herz, A., Fraling, F., Niedner, I., Farber, G., 1967. Pharmacologically induced alterations of cortical and subcortical evoked potentials compared with physiological changes during the awake-sleep cycle in cats. Electroenceph. Clin. Neurophysiol., Suppl. 26, 164–176.

Huttenlocher, P.R., 1960. Effects of the state of arousal on click responses in the mesencephalic reticular formation. Electroenceph. Clin. Neurophysiol. 12, 819–827.

Jouvet, M., 1962. Recherches sur les structures nerveuses et les mecanismes responsables de differentes phases du sommeil physiologique. Arch. Ital. Biol. 100, 125–206.

Kajimura, N., Uchiyama, M., Takayama, Y., Uchida, S., Uema, T., Kato, M., et al., 1999. Activity of midbrain reticular formation and neocortex during the progression of human non-rapid eye movement sleep. J. Neurosci. 19, 10065–10073.

Kakigi, R., Naka, D., Okusa, T., Wang, X., Inui, K., Qiu, Y., et al., 2003. Sensory perception during sleep in humans: a magnetoencephalograhic study. Sleep. Med. 4, 493–507.

Kasamatsu, T., Kiyono, S., Iwama, K., 1967. Electrical activities of the visual cortex in chronically blinded cats. Tohoku. J. Exp. Med. 93, 139–152.

Kräuchi, K., Knoblauch, V., Wirz-Justice, A., Cajochen, C., 2006. Challenging the sleep homeostat does not influence the thermoregulatory system in men: evidence from a nap vs. sleep-deprivation study. Am. J. Physiol. Regul. Integr. Comp. Physiol. 290, 1052–1061.

Lee, H.S., Ahn, J.H., Chung, J.W., Lee, K.S., 2008. Auditory steady-state response with the click auditory brainstem response in infants. Clin. Exp. Otorhinolaryngol. 1 (4), 184–188.

Loewy, D.H., Campbell, K.B., Bastien, C., 1996. The mismatch negativity to frequency deviant stimuli during natural sleep. Electroenceph. Clin. Neurophysiol. 98, 493–501.

Loomis, A.L., Harvey, E.N., Hobart, G.A., 1938. Disturbance-patterns in sleep. J. Neurophysiol. 1, 413–430.

Maquet, P., Degueldre, C., Delfiore, G., Aerts, J., Peters, J.M., Luxen, A., et al., 1997. Functional neuroanatomy of human slow wave sleep. J. Neurosci. 17, 2807–2812.

Maquet, P.A.A., Virginie Sterpenich, V., Albouy, G., Dang-Vu, T., Desseilles, M., Boly, M., et al., 2005. Brain imaging on passing to sleep. In: Parmeggiani, P.L., Velluti, R. A. (Eds.), The Physiologic Nature of Sleep. Imperial College Press, London, pp. 123–138.

Mendel, M.I., Goldstein, R., 1971. Early components of the averaged electroencephalographic response to constant level clicks during all night sleep. J. Speech Hear. Res. 14, 829–840.

Näätänen, R., 1995. Attention and Brain Function. Erlbaum, Hillsdale, NJ.

Nielsen-Bohlman, L., Knight, R.T., Woods, D.L., Woodward, K., 1991. Differential auditory processing continues during sleep. Electroenceph. Clin. Neurophysiol. 79, 281–290.

Ogilvie, R.D., Simons, I.A., Kuderian, R.H., MacDonald, T., Rustenburg, J., 1991. Behavioral, event-related potential, and EEG/FFT changes at sleep onset. Psychophysiology 28, 54–64.

Ornitz, E.M., Panman, L.M., Walter, R.D., 1967. The variability of the auditory averaged evoked responses during sleep and dreaming in children and adults. Electroenceph. Clin. Neurophysiol. 22, 514–524.

Osterhammel, P., Shallop, J., Terkildsen, K., 1985. The effects of sleep on the auditory brainstem response (ABR) and the middle latency response (MLR). Scand. Audiol. 14, 47–50.

Oswald, I., Taylor, A.M., Treisman, M., 1960. Discriminative responses to stimulation during human sleep. Brain 83, 440–453.

Pedemonte, M., Velluti, R.A., 2005a. What individual neurons tell us about encoding and sensory processing in sleep. In: Parmeggiani, P.L., Velluti, R.A. (Eds.), The Physiologic Nature of Sleep. Imperial College Press, London, pp. 489–508.

Pedemonte, M., Velluti, R.A., 2005b. Sleep hippocampal theta rhythm and sensory processing. In: Lander, M., Cardinali, D.P., Perumal, P. (Eds.), Sleep and Sleep Disorders: A Neuropsychopharmacological Approach. TX/Springer, New York, pp. 8–12. Landes Biosciencies.

Pedemonte, M., Peña, J.L., Torterolo, P., Velluti, R.A., 1996. Auditory deprivation modifies sleep in guinea pig. Neurosci. Lett. 223, 1–4.

Pedemonte, M., Drexler, D.G., Velluti, R.A., 2004. Cochlear microphonic changes after noise exposure and gentamicin administration during sleep and waking. Hear. Res. 194, 25–30.

Pedemonte, M., Medina-Ferret, E., Velluti, R.A., 2016. Sensory processing in sleep: an approach from animal to human data. In: Perumal, S.R.P. (Ed.), Synopsis of Sleep Medicine. APP. CRS, London, pp. 379–396. Chapter 22.

Peña, J.L., Pérez-Perera, L., Bouvier, M., Velluti, R.A., 1999. Sleep and wakefulness modulation of the neuronal firing in the auditory cortex of the guinea pig. Brain Res. 816, 463–470.

Perrin, F., García-Larrea, L., Mauguière, F., Bastuji, H., 1999. A differential brain response to the subject's own name persist during sleep. Clin. Neurophysiol. 110, 2153–2164.

Petrek, J., Golda, V., Lisonek, P., 1968. Cortical response amplitude changes produced by rhythmic acoustic stimulation in cats. Exp. Brain. Res. 6, 19–31.

Picton, T.W., Hillyard, S.A., Krausz, H.I., Galambos, R., 1974. Human auditory evoked potentials. I. Evaluation of components. Electroenceph. Clin. Neurophysiol. 36, 179–190.

Portas, C.M., 2005. Cognitive aspects of sleep. In: Parmeggiani, P.L., Velluti, R.A. (Eds.), The Physiologic Nature of Sleep. Imperial College Press, London, pp. 535–569.

Portas, C.M., Krakow, K., Allen, P., Josephs, O., Armony, J.L., Frith, C.D., 2000. Auditory processing across the sleep-wake cycle: simultaneous EEG and fMRI monitoring in humans. Neuron. 2, 991–999.

Pradham, S.N., Galambos, R., 1963. Some effects of anesthetics on the evoked responses in the auditory cortex of cats. J. Pharmacol. Exp. Ther. 139, 97–106.

Tanaka, H., Fujita, N., Takanashi, M., Hirabuki, N., Yoshimura, H., Abe, K., et al., 2003. Effect of stage 1 sleep on auditory cortex during pure tone stimulation: evaluation by functional magnetic resonance imaging with simultaneous EEG monitoring. Am. J. Neuroradiol. 24, 1982–1988.

Teas, D.C., Kiang, N.Y.S., 1964. Evoked responses from the auditory cortex. Exp. Neurol. 10, 91–119.

Ujszászi, J., Halász, P., 1986. Late component variants of single auditory evoked responses during NREM sleep stage 2 in man. Electroenceph. Clin. Neurophysiol. 64, 260–268.

Vanzulli, A., Bogacz, J., García-Austt, E., 1961. Evoked responses in man. III. Auditory response. Acta Neurol. Latinoamer. 7, 303–308.

Velluti, R.A., 1997. Interactions between sleep and sensory physiology. A review. J. Sleep. Res. 6, 61–77.

Velluti, R.A., Pedemonte, M., 2002. In vivo approach to the cellular mechanisms for sensory processing in sleep and wakefulness. Cell. Mol. Neurobiol. 22, 501–516.

Velluti, R.A., Pedemonte, M., García-Austt, E., 1989. Correlative changes of auditory nerve and microphonic potentials throughout sleep. Hear. Res. 39, 203–208.

Velluti, R.A., Pedemonte, M., Suárez, H., Bentancor, C., Rodriguez-Servetti, Z., 2010. Auditory input modulates sleep: an intra-cochlear-implanted human model. J. Sleep. Res. 19, 585–590.

Vital-Durand, F., Michel, F., 1971. Effets de la desafferentation periphérique sur le cicle veille-sommeil chez le chat. Arch. Ital. Biol. 109, 166–186.

Weitzman, E.D., Kremen, H., 1965. Auditory evoked responses during different stages of sleep in man. Electroenceph. Clin. Neurophysiol. 18, 65–70.

Wesensten, N.J., Badia, P., 1988. The P300 component in sleep. Physiol. Behav. 44, 215–220.

Wickelgren, W.O., 1968. Effects of the state of arousal on click-evoked responses in cats. J. Neurophysiol. 31, 757–768.

Williams, H.L., Tepas, D.I., Morloch, H.C., 1963. Evoked responses to clicks and electroecephalographic stages of sleep in man. Science 138, 685–686.

Zubek, J.P., 1969. Physiological and biochemical effects. In: Zubek, J.P. (Ed.), Sensory Deprivation: Fifteen Years of Research. Appleton-Century-Crofts, New York, pp. 255–288.

CHAPTER 5

Auditory Unit Activity in Sleep

The processing of sensory information is definitely present during sleep, although profound modifications occur. All sensory systems, visual, auditory, vestibular, somesthetic, and olfactory demonstrate some influence on sleep, and at the same time, sensory systems undergo changes that depend on the sleep or waking state of the brain (Velluti, 1997). Thus, not only different sensory modalities encoded by their specific receptors and pathways may alter the sleep and waking physiology, but also the sleeping brain imposes 'rules' on the incoming information. It is suggested that the neural networks responsible for sleep and waking control are actively modulated by the sensory input in order to enter and maintain sleep and wakefulness. Furthermore, both sensory stimulation and sensory deprivation may induce changes in sleep/waking neural networks. This leads to the conclusion that the central nervous system and sensory input have reciprocal interactions, on which normal sleep/waking cycling and behavior depend.

The present appraisal examines the results related to how auditory sensory information is worked out by the waking and sleeping brain from a single unit viewpoint. The auditory neuron firing rate, their temporal discharge distribution (pattern) and the relationship to hippocampus theta rhythm are considered components of auditory information processing (Velluti and Pedemonte, 2002; Pedemonte and Velluti, 2005; Velluti, 2005). Furthermore, considering auditory sensory system activity as an important piece of the sleep neurophysiology enigma, the influence of sensory incoming information is postulated as an active contributor intended for sleep processes development.

THE ACTIVITY OF CENTRAL AUDITORY NEURONS IN SLEEP AND WAKEFULNESS

In hearing, as well as in other sensory modalities, different types of mechanisms control the sensory input: (1) Prereceptorial actions regulate

The Auditory System in Sleep
DOI: https://doi.org/10.1016/B978-0-12-810476-7.00005-1

the stimulus energy through voluntary or reflex motor activities. Head and animal's pinna movements and contractions of the stapedius and tensor tympani muscles are examples of these regulating actions. (2) CNS actions provide control the efferent system. The central influence begins to be exerted at the receptor itself, as demonstrated by the changes imposed on the auditory nerve compound action potential and the cochlear microphonic recordings while the guinea pig is passing into sleep (see Chapter 1: Brief Analysis of the Auditory System Organization and Its Physiologic Basis).

The technical approach to reach single unit recordings in the sleep-waking physiological cycle is exhibited in the Fig. 5.1.

Sleep changes the neuronal activity in brainstem auditory nuclei and primary cortical loci. Fig. 5.2 presents percentages of firing changes of auditory neurons, recorded from partially restrained guinea pigs with glass micropipettes, during wakefulness (W) and sleep stages. The extracellular unitary response in each nucleus and in the primary cortex was grouped according to the firing rate changes upon passing to sleep. Neurons could

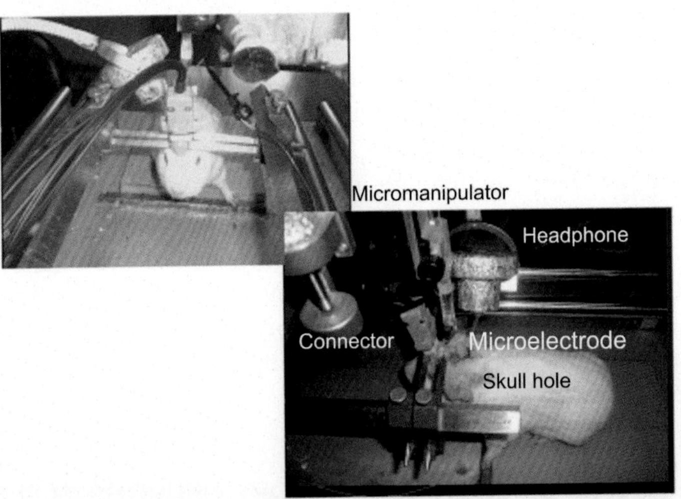

Figure 5.1 Guinea pigs recorded partially restricted with two metal bars attached to the skull. After recovery from surgery, these bars permit the animal head to be repositioned allowing the stereotaxic introduction of the glass micropipettes. Moreover, the body is placed in a hammock thus avoiding the head movements while allowing legs movements.

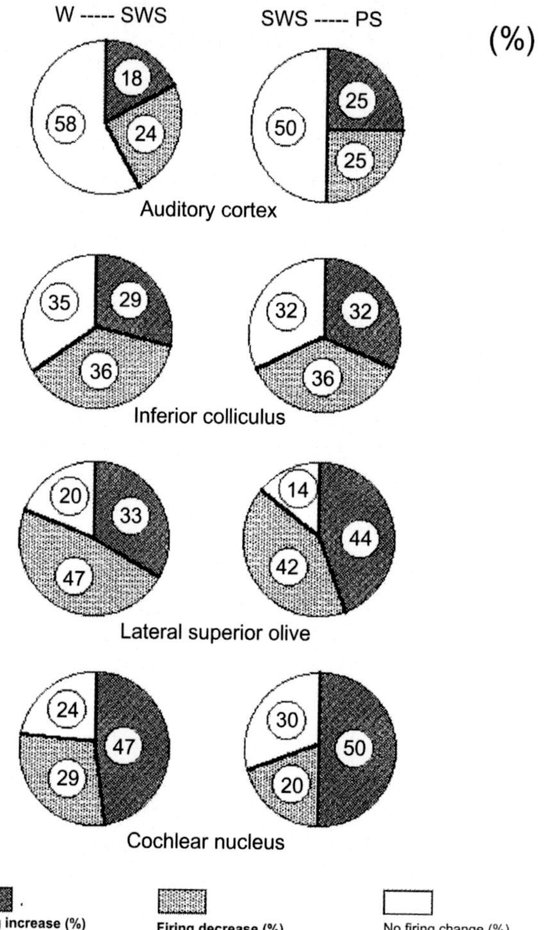

Figure 5.2 Guinea pig percentages of unitary evoked activity along the auditory pathway in the sleep-waking cycle. Pie charts show percentages (%) of neuronal firing shifts on passing from wakefulness to slow wave sleep and from slow wave sleep to PS. The subcortical nuclei, inferior colliculus, and the lateral superior olive exhibited a higher percentage of increasing-decreasing firing neurons. The cochlear nucleus presented increased activity, perhaps according to the higher blood flow reported by Reivich (1974). Around 50% of the cortical neurons responded as during wakefulness. No silent auditory neuron was detected on passing into sleep or during sleep in any pathway level. *Data from Peña, J.L., Pedemonte, M., Ribeiro, M.F., Velluti, R., 1992. Single unit activity in the Guinea-Pig cochlear nucleus during sleep and wakefulness. Arch. Ital. Biol. 130:179–189; Peña, J.L., Pérez-Perera, L., Bouvier, M., Velluti, R.A., 1999. Sleep and Wakefulness Modulation of the Neuronal Firing in the Auditory Cortex of the Guinea-Pig. Brain Res. 816:463–470; Pedemonte, M., Peña, J.L., Morales-Cobas, G., Velluti, R.A., 1994. Effects of sleep on the responses of single cells in the lateral superior olive. Arch. Ital. Biol. 132:165–178; Morales-Cobas, G., Ferreira, M.I., Velluti, R.A., 1995. Sleep and waking firing of inferior colliculus neurons in response to low frequency sound stimulation. J. Sleep Res. 4:242–251.*

decrease, increase, or exhibit no significant shifts in their firing in comparison to W. However, no completely silent neurons were detected when entering into sleep, slow wave sleep (SWS), or paradoxical sleep (PS) (Peña et al., 1992, 1999; Pedemonte et al., 1994, 2001; Morales-Cobas et al., 1995; Velluti et al., 2000; Velluti and Pedemonte, 2002; Pedemonte et al., 2004; Pedemonte and Velluti, 2005; Velluti, 1997, 2005). Therefore, it was concluded that the afferent input to the primary auditory cortex is not interrupted during any sleep phase; i.e., there is no auditory cortical functional deafferentation (Peña et al., 1999; Issa and Wang, 2008; Nir, 2016).

Approximately one half of the auditory cortical neurons responding to a characteristic frequency tone burst during sleep maintained a firing rate equal to that observed in a previous or subsequent W. The other ~50% was divided into neurons that: increased or decreased their firing rate. This second set of neurons (those that increased or decreased) may be part of neuronal networks that, in some unidentified way, could actively participate in sleep processes. Furthermore, I support the notion that the incoming sensory information is functionally active in sleep processes, not only as a passive participant (Velluti, 2005), this being a new approach for potential research.

The effects of sleep and wakefulness on auditory evoked activity of cats' single brainstem neurons, i.e., units at the mesencephalic reticular formation, was reported by Huttenlocher (1960). The activity of the non-lemniscal neuronal auditory pathway was shown to vary between sleep and W. The evoked activity of the units (about 50%) increased during quiet W and diminished during SWS. Approximately 30% of the recorded neurons showed evoked responses during SWS with the same activity, or even greater, than that during W. The most constant effect on evoked neuronal firing, as well as on spontaneous activity, was observed during PS in cats. During this sleep phase the responses to clicks diminished in the cells studied ($n = 16$) while in some neurons ($n = 5$) the activity evoked by the click was completely absent. The firing returned on arousal (Huttenlocher, 1960).

Four examples of changes in the unitary response on passing from W to SWS and PS are shown in Fig. 5.3.

A cochlear nucleus example neuron that exhibits a 'primarylike' post-stimulus time histogram increased its firing during SWS and PS. The control W periods, pre- and postsleep epochs, showed a remarkable W pattern and firing similarities.

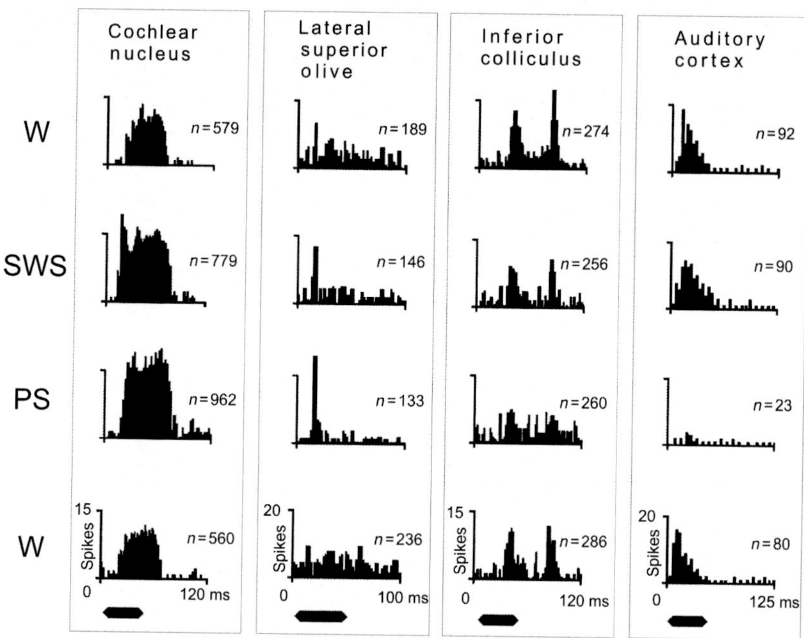

Figure 5.3 Four representative auditory neurons recorded at different auditory loci (cochlear nucleus, lateral superior olive, inferior colliculus, and auditory cortex) during the sleep-waking cycle are shown. Poststimulus time histograms exhibit changes in the pattern or frequency of discharge on passing from wakefulness (W) to slow wave sleep (SWS) and paradoxical sleep (PS). In these examples, the cochlear nucleus recorded neuron increases the firing rate during sleep maintaining the same pattern of discharge; the lateral superior olive shows both a change in pattern and a decrease in firing during sleep. The inferior colliculus neuron exhibits a changed pattern but not significant variation in firing rate. The auditory cortex neuron significantly decreases its firing only during PS, recovering it in the following W epoch. Stimuli: tone burst (50 ms, 5 ms rise-decay time, at the unit characteristic frequency). *Data from Peña, J.L., Pedemonte, M., Ribeiro, M.F., Velluti, R., 1992. Single unit activity in the Guinea-Pig cochlear nucleus during sleep and wakefulness. Arch. Ital. Biol. 130:179–189; Peña, J.L., Pérez-Perera, L., Bouvier, M., Velluti, R.A., 1999. Sleep and wakefulness modulation of the neuronal Firing in the auditory cortex of the Guinea-Pig. Brain Res. 816:463–470; Pedemonte, M., Peña, J.L., Morales-Cobas, G., Velluti, R.A., 1994. Effects of sleep on the responses of single cells in the lateral superior olive. Arch. Ital. Biol., 132: 165–178; Morales-Cobas, G., Ferreira, M.I., Velluti, R.A., 1995. Sleep and waking firing of inferior colliculus neurons in response to low frequency sound stimulation. J. Sleep Res. 4:242–251.*

In the anteroventral cochlear nucleus of the guinea pig extracellularly recorded responses revealed neurons having clearcut relationships with the sleep-waking cycle (Velluti et al., 1990; Peña et al., 1992). In agreement with previous results yielding increments in the averaged auditory nerve compound action potential amplitudes during SWS (Velluti et al., 1989), 72% of the cochlear nucleus units spontaneously discharging and 47% of the units responding to sound increased their firing in SWS. Some units (27%) showed no firing change as compared with W. During PS, 20% of sound responding and 66% of spontaneously firing neurons followed the auditory nerve amplitude trend. while on the other hand, a great proportion of units failed to follow the auditory nerve pattern. Differences in the probability of discharge over time (pattern) between the sleep phases and W were also observed (Fig. 5.4).

It has been postulated that the auditory efferent system modulates the auditory input at the level of the cochlear nucleus during sleep and W. The probability of firing and the changes in the pattern of discharge are thus dependent on the auditory input to both the cochlear nucleus and the brain functional state, asleep or awake (Fig. 5.4; Peña et al., 1992). Direct descending connections from cortical loci were also reported (see Chapter 1: Brief Analysis of the Auditory System Organization and Its Physiologic Basis).

The lateral superior olive example presents a different situation: The example unit progressively decreases its discharge rate when passing into sleep phases while the pattern of discharge changes neatly (Fig. 5.3). The neuron that initially, during W, was an 'on-sustained' unit is transformed into an 'onset' one during PS, i.e., discharging only at the beginning of the stimulus. Most neurons from the guinea pig lateral superior olive showed firing rate modulation on passing from W to SWS; 80% of the recorded cells changed their firing during binaural stimulation, while 85% did so during ipsilateral sound stimulation. In addition, shifts in the discharge pattern were observed in 15% of the cells recorded on passing from W into sleep. The most striking change was observed associated to the decreasing firing units: A very low discharge number, after a peak of responses in PS, during the last 40 ms, is shown in the poststimulus time histogram (Fig. 5.3).

The waking cues for binaural directional detection (in this particular experimental paradigm) disappeared during SWS; one possible interpretation of this result is that the binaural function of some (11.5%) lateral superior olive cells is impaired during SWS (Fig. 5.5).

Auditory efferent pathways are impinging on the lateral superior olive neurons, thus, their activity is dependent on ascending input as well as on cortical descending influences, the latter being behaviorally dependent, sleep and waking in our case (Pedemonte et al., 1994).

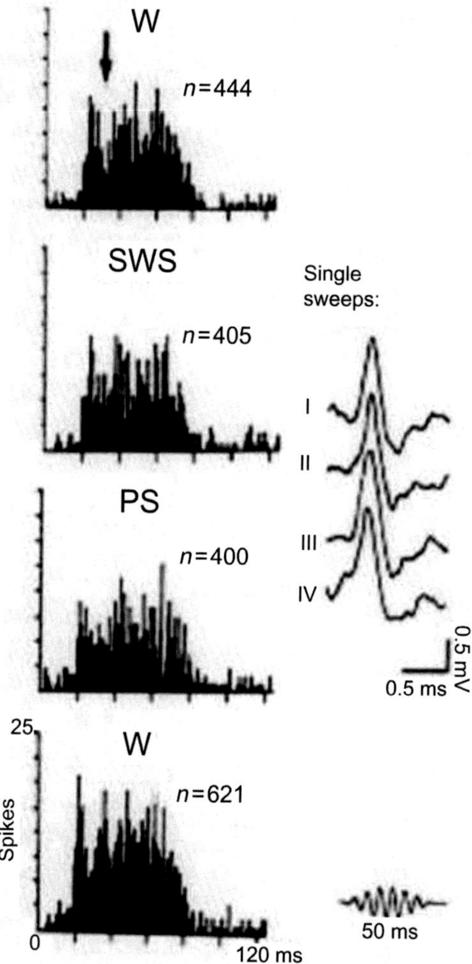

Figure 5.4 Qualitative changes in probability of discharge over time (pattern) of a cochlear ventral nucleus auditory neuron during wakefulness (W) and sleep stages. The upper W epoch shows a dip of decreased firing probability (arrow) in the post-stimulus time histogram. During the successive episodes of slow wave sleep (SWS) and paradoxical sleep (PS), neither the firing nor the pattern change significantly although with a different probability: during W the firing dip disappears. The last W epoch (after sleep) shows both a discharge increase ($n = 621$) and a pattern shift, exhibiting a 'chopper' type of response, a quite different pattern. On the right side, individual sweeps are shown (I, II, III, IV) demonstrating the same unit was always recorded. Unit characteristic frequency: 0.85 kHz; 48 dB SPL. *Modified from Peña, J.L., Pedemonte, M., Ribeiro, M.F., Velluti, R., 1992. Single unit activity in the Guinea-Pig cochlear nucleus during sleep and wakefulness. Arch. Ital. Biol. 130:179–189.*

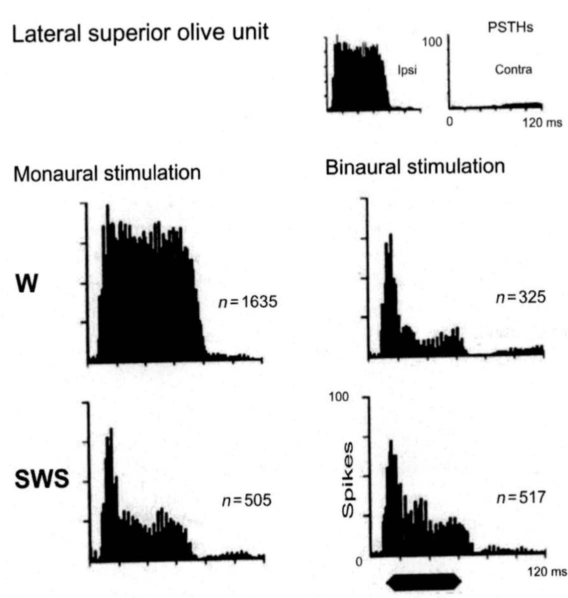

Figure 5.5 Tone-evoked responses from a lateral superior olive neuron during waking (W) and slow wave sleep (SWS), with monaural and binaural sound stimulation. Upper traces show the unit response to ipsi- and contralateral auditory stimuli (45 dB SPL; characteristic frequency 1.0 kHz) exhibiting neat differences. During W, monaural and binaural sound stimulation showed great changes, including pattern and firing probability ($n = 1635$ vs $n = 325$). On passing to SWS the neuron exhibited lower firing ($n = 505$) with monaural stimulation, the same pattern and a very similar discharge number with binaural stimulation; this neuron and its neuronal network cannot recognize the difference between monaural and binaural sound stimulation in this sleep stage, SWS. *Modified from Pedemonte, M., Peña, J.L., Morales-Cobas, G., Velluti, R.A., 1994. Effects of Sleep on the Responses of Single Cells in the Lateral Superior Olive. Arch. Ital. Biol. 132: 165–178.*

The example neuron of the inferior colliculus changes the pattern, i.e., the temporal distribution of the neuronal discharge, in spite of a non-significant firing number shift, on passing from W to sleep phases (Fig. 5.3). Most neurons (63%) exhibit evoked firing rate increases or decreases on passing from W to sleep. The majority of the inferior colliculus units shift their evoked firing during PS, while only 11% did not show such changes. Moreover, inferior colliculus cells spontaneous discharge is observed to increase in most units during PS (Morales-Cobas et al., 1995). Auditory information processing is present during sleep stages although with differences regarding W (Fig. 5.6).

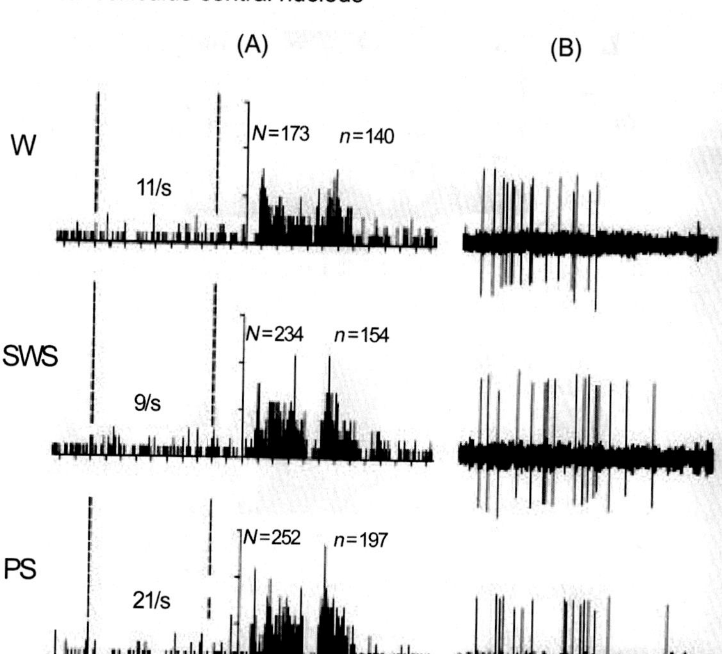

Figure 5.6 Evolution of an inferior colliculus auditory unit firing throughout the sleep-waking cycle. In A, the peristimulus time histograms (PSTHs) exhibited an increased evoked firing (N) on passing from wakefulness (W) to slow wave sleep (SWS) and paradoxical sleep (PS). The second half of the PSTH (*n*) showed increasing firing on passing from W to SWS, while when changing from SWS to PS, another discharge decrease was present. In B, a unit raw recording is shown exhibiting the split discharge pattern observed in the PSTHs. On the other hand, the spontaneous firing rate (PSTH left side) presented a slight firing decrease on passing from W to SWS, while a highly significant ($p < .001$) increase was obtained during PS. *Modified from Morales-Cobas, G., Ferreira, M.I., Velluti, R.A., 1995. Sleep and waking firing of inferior colliculus neurons in response to low frequency sound stimulation. J. Sleep Res. 4:242–251.*

The inferior colliculus auditory neurons send descending connections to regions such as the dorsal pontine nuclei, postulated to mediate sleep signs and processes, making this locus suitable for sleep-auditory system interactions. In addition, both the cochlear and trapezoid nuclei are sources of abundant projections to the rostral part of the nucleus *reticularis pontis caudalis* (Garzón, 1996), which in turn is strongly connected with the PS sign-induction zone located in the ventral part of the nucleus *reticularis pontis oralis* (Reinoso-Suárez et al., 1994).

Neurons of the medial geniculate nucleus, the auditory thalamus, showed evoked firing shifts mainly decreasing on passing from W into SWS, while the spatial receptive field was preserved, indicating that the information sent to cortical cells may carry significant content (Edeline et al., 2000; Edeline, 2003).

The auditory areas of the cerebral cortex receive complex ascending input, originating from both ears, and, in turn, send projections to thalamic and midbrain targets (Saldaña et al., 1996). Thus, the auditory cortex may control the whole auditory system and, at the same time, be dependent on the general state of the brain, asleep or awake (Nir, 2016).

In the guinea pig, auditory cortical units recorded at their characteristic frequency varied their firing evoked activity on passing from W into sleep phases. The firing shifts were observed during SWS, exhibiting decreased-increased firing while a percentage did not show firing rate changes (Fig. 5.2). Extracellular recordings of both spontaneous and sound evoked responses from the primary auditory cortex (A1) have been shown to be highly dependent on the state of the brain, W, SWS, or PS (Fig. 5.3). Thus, the sleep and W brain states modify the auditory cortical processing of simple sound stimuli (Peña et al., 1999). Issa and Wang (2008) found similar results in primates.

An important result in the Peña et al. (1999) study is the divergent change obtained between spontaneous and evoked unitary activity. The trend was not always the same; the evoked activity could increase while the spontaneous activity decreased or vice versa. A hypothesis that emerges from these results is that at least two separate controls are exerted on the same neuron: one correlated with the incoming signals and a second control perhaps from a nonauditory CNS locus and from stored auditory information loci. As reported in the primate visual cortex, the secondary visual areas increased their activity during PS when studied with positron emission tomography (Braun et al., 1997, 1998). Two questions readily arise: Was the neuronal activity in those visual areas related to the visual imagery normally occurring during PS? And was the increased spontaneous activity in our auditory cortical results, particularly during PS, also related to auditory hypnic 'images'?

It is remarkable that in our results a number of the neurons in the primary auditory cortex exhibited no significant quantitative changes in their evoked or spontaneous firing rates that could be correlated to behavioral, brain state shifts. This is inconsistent with reports in the literature but consistent with the hypothesis that, to a varying extent, the responsiveness of the auditory system is preserved during sleep. However, an analysis in which the temporal development of the neuronal discharge is studied over the sleep-wakefulness cycle remains to be done. Nevertheless, the present study demonstrates that significant changes can be correlated to behavior during either spontaneous or evoked activity in 42.2%–58.3% of the sampled neurons. Such auditory cortical neurons (and the neural networks to which they belong) exhibit a different behavioral state-related functioning, probably reflecting a general shift in the brain sensory processing during sleep. About half of the auditory cells were as active as during W on passing into SWS or PS. This cortical region is not disconnected from the outer world; on the contrary, it is, perhaps monitoring sounds from the environment.

We postulated more than once that thalamocortical 'blockade' of information in sleep was not real (Peña et al., 1999; Velluti and Pedemonte, 2005, 2012). This approach was recently supported, signaling that sensory relay remains functional during sleep (Sela et al., 2016) and that sensory disconnection in sleep is not a real phenomenon.

Neuronal Discharge Pattern Shifts

The firing pattern change may support a different possibility of sound analysis as well as suggest a different mode of relation to other cell assemblies or networks, which I postulate as related to active sleep processes. The same neuron may exhibit one pattern during SWS and a different one during PS, to recover the initial firing distribution at the following W epoch. Moreover, diverse patterns could be observed throughout the sleep-waking cycle (Velluti and Pedemonte, 2002), perhaps in association to another, different neuronal network. Besides, to further demonstrate how the auditory neuronal firing changes at the cortical level on passing from W into SWS and PS and back from PS to W, Fig. 5.7 shows this dynamic condition during a full sleep cycle.

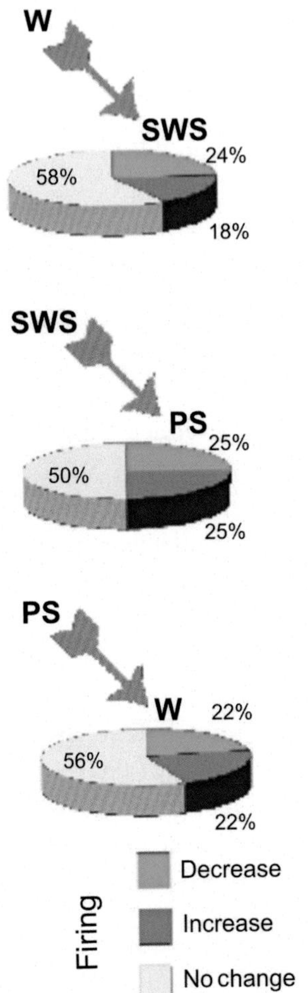

Figure 5.7 The pie charts show the firing evolution of an auditory cortical neuron when passing from wakefulness (W) to sleep and again into W.

This is indicative of a neatly reversible physiologic process, being also another experimental demonstration that the afferent input to the auditory cortex is dynamic, although predictable, and that the auditory cortex is not functionally deafferented during sleep. In the primary auditory cortex (AI) example, a continuously changing unitary firing is shown on passing from W to SWS, from SWS to PS and from PS to W. The firing recovery in the subsequent W after a PS epoch once a sleep cycle is completed, exhibits the relative constancy of the phenomenon.

Bursting units firing at the auditory cortical level are not, or in very few cases, present during W or sleep. The proportion of such bursting neurones in anaesthetized states is much larger (Fig. 5.8), i.e., it gives the impression of an artefact phenomenon rather than a physiologic one. In my experimental data it was never recorded a bursting cell at the cortical level during W or sleep in guinea pigs.

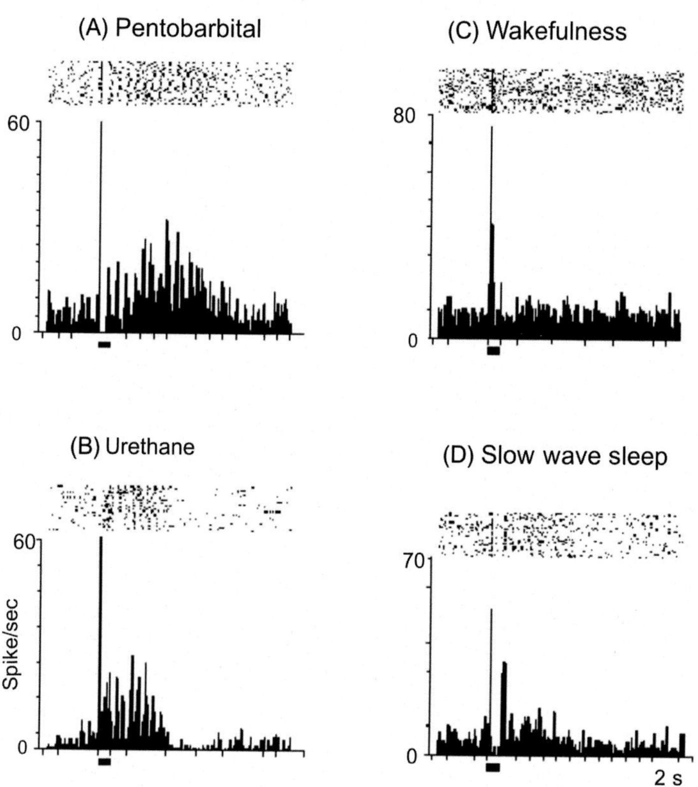

Figure 5.8 Tone-evoked oscillations are present in the poststimulus time histograms (PSTHs) of anesthetized animals but not in nonanesthetized, awake or asleep animals. Although the PSTHs and rasters exhibit oscillations under pentobarbital or urethane anesthesia, there is no tendency for such oscillating pattern in waking or slow wave sleep. *Modified from Edeline, J.-M., 2005. Learning-induced plasticity in the thalamo-cortical auditory system: should we move from rate to temporal code descriptions? In: R. Konig, P. Heil, E. Budinger, H. Scheich (Eds.), The Auditory Cortex. Lawrence Erlbaun Ass., Mahwah, New Jersey, London. pp 365–382 (Edeline, 2005).*

Brain's Math

Neural models use mathematical operations to predict brain function. How realistic such operations are is a matter of debate. The auditory system offers a useful substrate to approach this question because it is possible to test specific mathematical operations by manipulation of acoustic signals.

In order to determine the identity and location of acoustic objects, the brain integrates the information arriving to each ear, separating signal from noise. Comparing and combining bilateral signals and removing noise can be represented by equations. Cross-correlation and multiplication are mathematical correlates of comparison and combination selectivity, respectively. Likewise, a common mathematical tool to extract signal from noise is averaging across repetitions. In the auditory pathway, we can find neurons that perform each one of these types of processing.

Neurons in the auditory brain stem are able to detect minute differences in time between the signals at the left and right ears, a major cue for sound localization. This ability can be described as a running cross-correlation of signals from each ear, a computation that allows neurons to indicate the degree of similarity between the right- and left-side inputs in their firing rate. The space-specific neurons that constitute the map of auditory space in the avian midbrain are selective to combinations of spatial cues. This combination is implemented by an effective multiplication of inputs from independent processing pathways. Multiplication not only makes the response to particular combinations robust across a broad range of sound levels but also lends veto power to one spatial cue over another. In the central nucleus of the inferior colliculus, neurons encode sound direction more reliably than in earlier stages of the pathway, through an averagelike processing. Thus, the hierarchically organized auditory system implements computations that models predict, lending support to the notion that mathematical operations are a powerful description of brain function.

José L. Peña
Dominick P Purpura Department of Neuroscience,
Albert Einstein College of Medicine, New York, USA

RELATIONSHIPS BETWEEN THE THETA RHYTHM OF THE HIPPOCAMPUS AND AUDITORY NEURONS

The hippocampus produces a high amplitude transient theta rhythm when, e.g., a cat is looking at itself in a mirror; this observation by Grastyán et al. (1959) connected this rhythm to a sensory input, the vision, and to higher processes as, perhaps, conspecific recognition

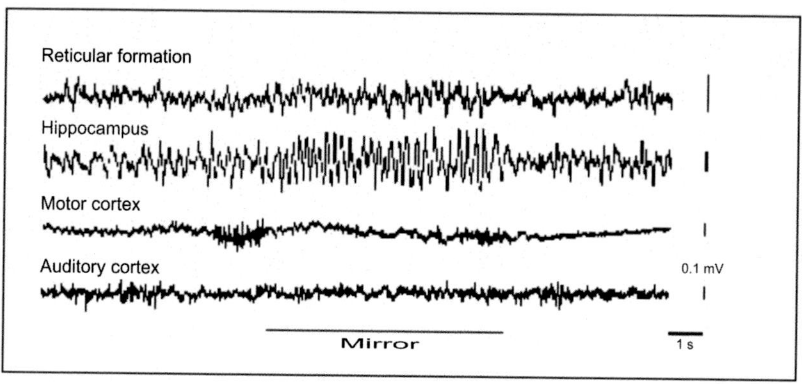

Figure 5.9 Recording from an awake cat showing enhancement of theta rhythm in the hippocampus when the animal sees itself at a mirror (bar). *Modified from Grastyán, E., Lissák, K., Madarász, I., 1959. Hippocampal activity during the development of conditioned reflex. Electroenceph. Clin. Neurophysiol. 11:409–430.*

(Fig. 5.9). At the same time, recordings carried out in the primary auditory cortex showed evoked neuronal firing shifts elicited by electrical stimulation of the hippocampus, indicating that these brain regions are interconnected and exhibit a functional relationship. These results support the notion that an auditory-hippocampus shared functional interaction, although unknown in detail, may be present (Cazard and Buser, 1963; Redding, 1967; Pompeiano, 1970; Parmeggiani and Rapisarda, 1969; Parmeggiani et al., 1982; Buszaki, 1996).

The ultradian hippocampal theta rhythm, within the wakefulness-sleep circadian cycle, may modulate sensory neuronal activity. It has been one of the conspicuous time givers postulated as an internal *zeitgeber*, a temporal organizer for auditory sensory processing (Pedemonte et al., 1996; Velluti et al., 2000; Velluti and Pedemonte, 2002; Pedemonte and Velluti, 2005). The theta waves, 4–10 cycles per second (cps), may affect spatially distant neurons by inducing fluctuations of the cellular excitability due to membrane potential oscillations (García-Austt, 1984; Kocsis and Vertes, 1992). Although more prominent in active W and PS, the theta rhythm is always present in the brain; e.g., the hippocampus theta frequencies can also be observed during SWS when analyzed in the frequency domain, the Fourier transform (Komisariuk, 1970; Gaztelu et al., 1994; Pedemonte et al., 1996).

This particular hippocampus rhythm has been related to several brain processes; it was found involved in motor activities during both W and PS (Buño and Velluti, 1977; García-Austt, 1984; Lerma and García-Austt,

1985) as well as in the sensory processing related to a motor context (Grastyán et al., 1959; Kramis et al., 1975). It was also found implicated in spatiotemporal learning (Winson, 1978; O'Keefe and Recce, 1993), associating distant, discontiguous events (Wallestein et al., 1998), and learning of temporal sequences (Metha et al., 2002). A role of the theta rhythm in learning and memory has been proposed from different viewpoints (Adey et al., 1960; Burgess and Gruzelier, 1997; Doppelmayr et al., 1998; Klimesch, 1999; Kahana et al., 1999; Vinogradova, 2001), and it has also be linked to the modulation of autonomic processes such as the heart rate in the guinea pig and human (Pedemonte et al., 1999, 2003).

Rythms and Auditory Neuronal Activity

One of the most exciting experiences of in vivo electrophysiology is to watch a neuron discharging online. The neuronal decision to fire, or not, an action potential can integrate information carried by thousands of synaptic inputs at a given time. Central auditory neurons can discharge tens of spikes per second, not all of them being synchronized with sound stimuli.

Because of the auditory unit firing differences observed between wakefulness and sleep, the question is what information do the neurons that shift firing in sleep carry? Perhaps, those units are related to some sleep processes.

Research in audition deals mainly with the study of neuronal responses to pure tones, white noise, and artificially synthesized acoustic signals. However, the real auditory world consists of complex combinations of frequencies, whose power is distributed along time in specific ways, and this world is interpreted by the brain according to its current state, waking or asleep. Getting to know how the brain processes that information is perhaps the most important challenge for neuroscientists.

The brain presents many spontaneous rhythms, which can be recorded as field potentials. These rhythms are the result of internal processes that change with states such as sleep, wakefulness, and attention. Therefore, I am speculating, brain rhythms must be involved in information processing under different conditions. Auditory information needs a temporal frame for events to be analyzed in particular sequences. Words and music have sense if they are perceived along time. Among the variety of rhythms that the brain offers, the theta rhythm (4–12 cps) could be a good candidate because (1) it is present in any behavioral state, (2) it is generated, amplified, and distributed by a structure involved in learning and memory processes, the hippocampus; and (3) it has been related to auditory and visual sensory systems, motor and autonomic processes during sleep and wakefulness. Furthermore, studying mice, it has been established that theta frequency in PS is controlled by a single

(*Continued*)

(Continued)

autosomal recessive gene, suggesting that kind of genes may be found for rhythms variant during human sleep.

Over more than a decade of research, our group found a temporal correlation between the hippocampal theta rhythm and the discharge of auditory neurons, postulating that the theta rhythm is a timer in auditory processing, which lends experimental support to the hypothesis of this rhythm as a time giver.

What happens with the human being? Many human functions are rhythmic, such as respiration, heart beating, automatic movements (walking, running, swimming, etc.). It is not only easier to learn rhythmic tasks but humans are attracted to sounds that have a particular *tempo*. Have you ever observed people's behavior in a music concert? You can see them performing enthusiastic rhythmic movements, often synchronized with frequencies within the delta and theta range. When music becomes slower or faster, people abandon the synchronized movements.

In conclusion, brain rhythms (which change during sleep and wakefulness) could offer different frequency ranges for the temporal, and perhaps also spatial, organization of the brain auditory processing.

Marisa Pedemonte
Universidad CLAEH, Punta del Este, Uruguay, 2008

The auditory neurons exhibit phase locking, i.e., temporal correlation with theta rhythm at different levels of the auditory pathway, from the brain stem (the cochlear nucleus, the superior olive, the inferior colliculus) to the primary cortical region (Pedemonte et al., 1996; Velluti et al., 2000; Velluti and Pedemonte, 2002; Pedemonte and Velluti, 2005). The processing in other sensory modalities has been associated to the theta rhythm, such as touch (Nuñez et al., 1991), pain (Vertes and Kocsis, 1997), vision (Gambini et al., 2002), and olfaction (Margrie and Schaefer, 2003; Affani and Cervino, 2005). Therefore, it is my tenet that the theta rhythm of the hippocampus contributes, in the time domain, to the complex dilemma of the central auditory information processing in sleep and waking.

Most lateral superior olive auditory neurons showed a spontaneous firing rate modulation on passing from wakefulness to slow wave and PS (Fig. 5.10; Velluti et al., 2000). The interactions between the auditory units and the hippocampal theta, demonstrated by the cross-correlations, exhibited phase locking during wakefulness and PS, at the same hippocampal theta wave phase. In this example, during SWS, the phase locking

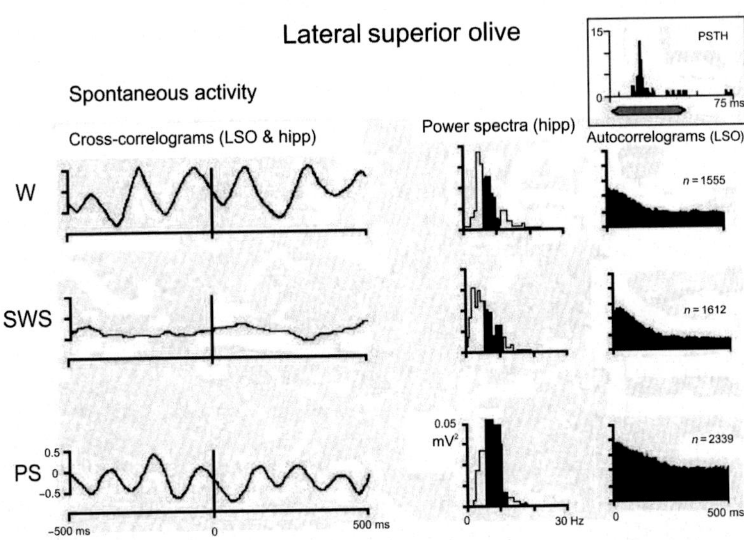

Figure 5.10 Functional relationship between a lateral superior olive neuronal spontaneous discharge and hippocampal (hipp) theta rhythm during behavioral states. The auditory unit firing is phase locked with the hipp theta rhythm during wakefulness (W) and paradoxical sleep (PS). In the example shown, the correlation is lost during slow wave sleep (SWS). On the other hand, the firing rate increases during SWS and particularly during PS (autocorrelogram). The power spectra exhibit a strong theta frequency band power in each state. Inset: poststimulus time histogram (PSTH) of the unit; binaural sound stimulation (tone burst at characteristic frequency: 2.0 kHz; 48 dB SPL). *Modified from Velluti, R.A., Pedemonte, M., Peña, J.L., 2000. Reciprocal actions between sensory signals and sleep. Biol. Signals Recept. 9:297–308.*

failed to appear in spite of a hippocampal theta occurrence shown in the corresponding power spectrum. In other unit recordings the phase locking was also present during SWS.

Inferior colliculus neurons showed a higher synchrony with hippocampal theta when sound stimulated at the unit's characteristic frequency during waking, although the spontaneous activity also exhibited phase locking (Fig. 5.11).

Shifts in the angle of phase locking to the theta rhythm were also observed during PS. Moreover, when a continuous pure tone (at the neuron characteristic frequency) was delivered, the single unit response became rhythmic at the theta frequency (Pedemonte et al., 1996). During PS all inferior colliculus auditory neurons recorded exhibited hippocampal theta correlation: 40% were rhythmic and phase locked to the theta frequency and 60% were nonrhythmic maintaining the theta phase locking (Fig. 5.12).

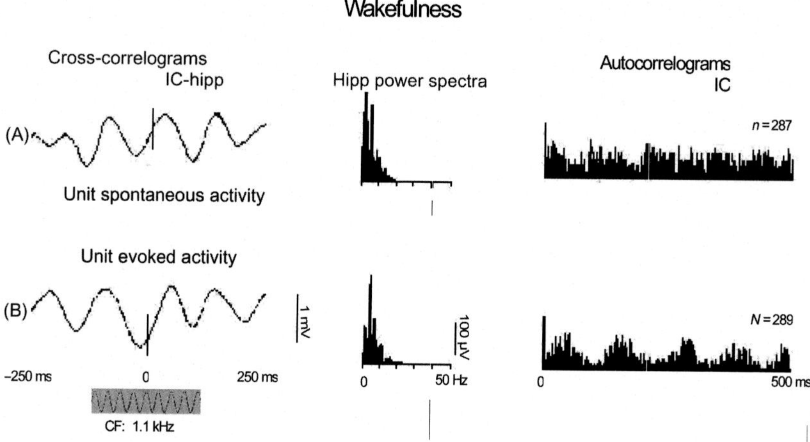

Figure 5.11 Temporal correlation between hippocampal (hipp) theta rhythm, and inferior colliculus (IC) auditory single unit spontaneous and evoked activity. Both, spontaneous and evoked firing showed cross-correlations (CC) (phase-locking) with hippocampal (bipolarly recorded) theta rhythm. The phase locking showed a small timing difference between the CC with spontaneous and evoked firing. The autocorrelation histograms (AC) exhibited that the nonrhythmic discharge pattern during spontaneous activity (upper trace) becomes rhythmic at theta frequency during the evoked activity (lower trace), thus stressing the relevant influence of the auditory input on the theta waves. The correlation of the spikes with the theta rhythm was studied with spike-triggered averaging of the hippocampus field activity. Continuous pure-tone sound stimulation at the unit's characteristic frequency (CF): 1.1 kHz; 10 dB above threshold. *Modified from Pedemonte, M., Peña, J.L., Velluti, R.A., 1996. Firing of inferior colliculus auditory neurone is phase-locked to the hippocampus theta rhythm during paradoxical sleep and waking. Exp. Brain Res. 112:41–46.*

The hippocampal formation is a high level cortex specifically involved in memory encoding. Moreover, it has been suggested that the hippocampal circuitry organization serves as a matrix for rapid autoassociative memory (McNaughton and Morris, 1987). Sounds such as a word or a musical phrase, develop in temporal sequences, so a special rapid memory and temporal organization for receiving the signals may be necessary. Furthermore, the hippocampus 'appears as a sort of holding system that is necessary for the temporary storage of information regarding the temporal order and the spatial context of events' (Lopez de Silva et al., 1990; Rawlins, 1985). Finally, the hippocampal theta rhythm may play this role as an internal clock (*zeitgeber*), adding a temporal dimension to the processing of auditory sensory information (Velluti and Pedemonte, 2002).

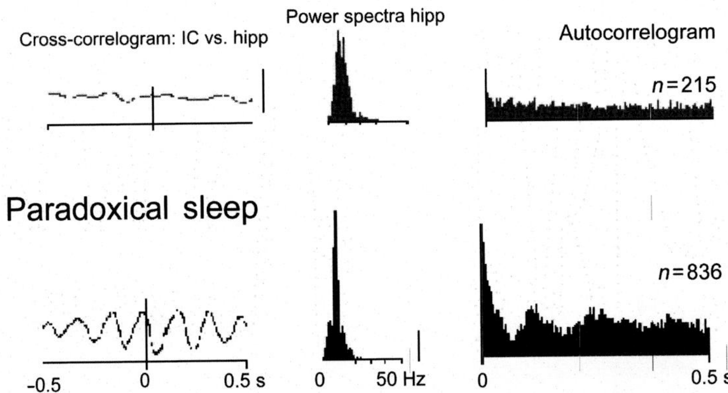

Figure 5.12 Cross-correlation of CA1 hippocampus theta rhythm and inferior colliculus auditory neuron spontaneous activity during wakefulness and paradoxical sleep. This neuron behaved in a nonrhythmic manner without phase locking to the theta rhythm during wakefulness. The autocorrelogram also shows this unit as nonrhythmical. In the paradoxical sleep stage, it spontaneously became both rhythmic and phase locked to the hippocampal theta rhythm and the unit firing increased. The theta rhythm frequency was enhanced around the 6–7c/s band. Cal, Crosscorrelogram; 1 mV; Power spectra bar, 50 μV. *Modified from Pedemonte, M., Peña, J.L., Velluti, R.A., 1996. Firing of inferior colliculus auditory neurone is phase-locked to the hippocampus theta rhythm during paradoxical sleep and waking. Exp. Brain Res. 112:41–46.*

Furthermore, the hippocampus is involved in the neural coding of spatial positions (O'Keefe and Recce, 1993; Wallestein et al., 1998; Best et al., 2001), and it is necessarily associated with sensory processing to establish its space location. As an experimental animal traverses space, the hippocampal place neurons firing progressively change to an earlier phase of the ongoing theta rhythm (Skaggs et al., 1996; Magee, 2003).

In agreement with the writer Jorge L. Borges (1899–1986), 'Movement, that is occupying different positions ... is unconceivable without time', time, not only space, is the other variable that may be controlled by the hippocampus and it is represented by the field activity, the theta rhythm, which is postulated as a meaningful factor in the temporal processing of auditory signals. The temporal correlation phenomenon (phase locking) was described during W as well as in both sleep phases, SWS and PS (Pedemonte et al., 1996, 2001; Velluti et al., 2000; Velluti and Pedemonte, 2002).

Phase locking is not a fixed phenomenon but may change due to known and unknown factors. For instance, when a sound appears or a different sensory stimulation is introduced to distract the animal's attention, it is possible to evoke the theta phase locking and the increment of theta power for many seconds (Fig. 5.13). These shifts occur most of the time in W as well as during all sleep stages in the inferior colliculus auditory recordings (Liberman et al., 2006). Furthermore, lateral geniculate visual neurons may evoke unitary discharge phase locking when changing the frequency of light flashes (Pedemonte et al., 2005). In

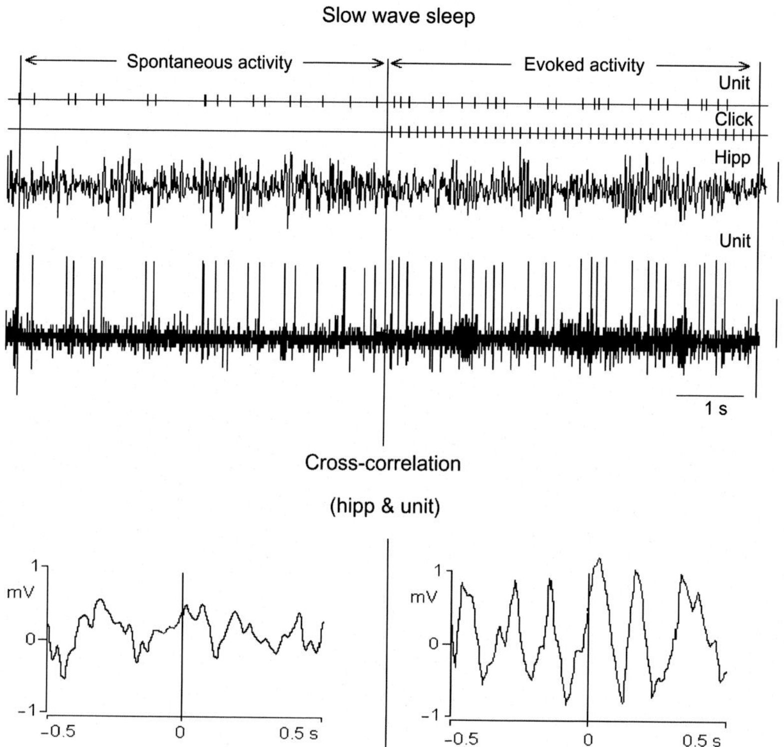

Figure 5.13 Relationship between hippocampal theta rhythm and auditory cortical unit during slow wave sleep. Upper traces: raw data showing, from top to bottom, digitized units, sound stimuli (clicks), hippocampal field electrogram (hipp), and auditory cortical unitary discharges. The spontaneous activity is shown on the left of the vertical line and, on the right, the activity during sound stimulation (8/s). The cross-correlogram did not exhibit phase locking with the hipp theta rhythm during the auditory unit spontaneous discharge. However, when the sound stimulation started, the neuron began to fire in close correlation with a particular theta rhythm phase, phase locked. Cal: hipp, 1 mV; unit, 50 μV. *Modified from Velluti, R.A., Pedemonte, M., Peña, J.L., 2000. Reciprocal actions between sensory signals and sleep. Biol. Signals Recept. 9:297–308.*

addition, a cyclic on and off temporal correlation to the theta rhythm of about 5 seconds has been observed in previous reported data similar to our results that change the attention level during ~5 seconds (Vinogradova, 2001; Velluti and Pedemonte, 2002). Vinogradova (2001) classifies the modulating influence of theta rhythm in two systems: (1) a regulatory system, linking the hippocampus to the brainstem structures, which senses the attention level, introducing primary information about changes in the environment, and (2) an informational system that holds reciprocal interactions with the neocortex (Fig. 5.14).

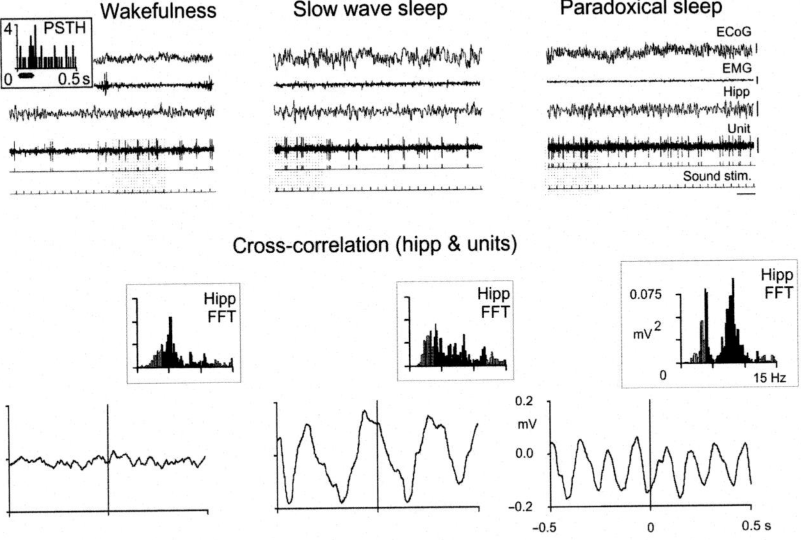

Figure 5.14 Temporal correlation (phase locking) between the evoked discharges of a primary auditory cortex neuron and hippocampus (Hipp) theta rhythm, during sleep-wakefulness cycle. Top, raw recordings of the electrocorticogram (ECoG), electromyogram (EMG), hippocampus field activity, and neuronal discharge (unit) during wakefulness, slow wave and paradoxical sleep. Digitized units and acoustic stimuli are shown below. The left corner inset exhibits the poststimulus time histogram (PSTH) in response to a pure tone-burst stimulus at the unit characteristic frequency. Bottom, the cross-correlation between hipp field activity and auditory units firing was obtained by spike-triggered averaging. The insets show the hipp power spectra (FFT) with the theta range in black. In this example, the neuronal discharge is phase locked with the hipp theta rhythm during slow wave and paradoxical sleep whereas no temporal correlation appears in a wakefulness epoch. Cals.: ECoG, 0.5 mV; EMG, 0.1 mV; Hipp, 0.5 mV; Unit, 0.1 mV; time, 1 s. *Modified from Velluti, R.A., Pedemonte, M., 2002. In Vivo approach to the cellular mechanisms for sensory processing in sleep and wakefulness. Cell Mol. Neurobiol. 22:501–516.*

Human intracranial recording has revealed theta oscillations in cortical places, suggesting the existence of theta generators in the brain surface. Theta waves were observed in the basal temporal lobe and frontal cortex without a functional coupling between neocortex and hippocampus during theta periods. This is indicative of multiple theta generators in human brain and evolving from tonic (in lower mammals) to phasic in PS (Kahana et al., 2001; Cantero et al., 2003).

Brain Plasticity Versus Brain Homeostasis

One of the most basic concepts underlying contemporary neuroscience is that the adult mammalian brain changes every time that a new motor or cognitive ability is acquired and stored. However, this basic tenet is supported by several assumptions that need to be stated clearly. First, as part of a changing universe, the brain is always changing. We, as neuroscientists, should distinguish which changes are spontaneous and which represent the substrate of learning acquisition and storage. Second, even when the brain changes either spontaneously or due to its own activity-dependent plasticity, humans are still capable of maintaining concepts, ideas, and beliefs for relatively long periods of time. Finally, the brain can also suffer significant structural changes in the absence of any noticeable behavioral or cognitive changes, as in the early asymptomatic periods of many neurodegenerative diseases or during some regenerative processes. Therefore, it seems the brain permanently faces a decision between plastic versus stabilizing processes.

The term *brain plasticity* is not necessarily synonymous with brain functioning. Regarding plasticity, we assume the presence of some structural, albeit localized and subtle, neural changes underlying learning and memory processes. But the brain is also able to be in different states (asleep, awake, or dreaming) involving changes like receiving sensory information that are not necessarily structural but rather functional. Thus, it is still possible that learning mechanisms are ascribed to the dynamic, emergent properties of neural ensembles. We have more neurons than (different) proteins, and perhaps the former can carry out a good job without the need of any structural modifications of their already sophisticated connectivity. Why, then, do most neuroscientists prefer to lean on neural plasticity rather than on neural functional states? The most parsimonious answer is that we have collected a huge amount of information about the structure and connectivity of neural tissue at subcellular and molecular levels and about the anatomical and biochemical rules and pathways maintaining these structures and circuits. In addition, definite behavior and sensory-motor properties are easily ascribed to specific neural sites. In contrast, our information about brain functioning during learning

(*Continued*)

(Continued)

situations is too constrained by the limitations imposed by electrical recordings from small numbers of neural elements selected out of billions or by modern mapping techniques dealing with electrical or biochemical representations of brain activity.

Although rerouting of given neural connections is proposed from time to time, none has been convincingly demonstrated in the brain of adult mammals. In fact, it has never been proven that postembryonic peripheral or central neurons are able to respecify their functional codes following their reinnervation of new neural or muscle targets. Plasticity is then restricted to the local microenvironment surrounding postsynaptic neurons, although molecular processes involving presynaptic terminals cannot be ruled out. These structural changes are normally described as the final steps of functional changes named *long-term potentiation* and *long-term depression*. Therefore, the search for the neural engram or learning trace (either localized at specific cortical or subcortical structures or properly distributed in the neural tissue) has now been replaced by the search for a (mostly intracellular) microengram. In accordance with this generally assumed, contemporary trend, selected brain sites should be endowed with the necessary and sufficient molecular machinery to produce the pre- and/or postsynaptic changes supporting the acquisition, storage, and retrieval of motor and cognitive learning.

José M. Delgado-García
Universidad Pablo de Olavide, Seville, Spain

Although a temporal correlation between the hipp theta and the auditory cortex unitary activity does not necessarily imply a direct influence of one over the other, the conspicuous correlation of hipp theta and several other neural processes suggests the existence of some kind of interaction. This correlation has been observed during W and also during both sleep phases. It is consistent with the proposed hypothesis that hipp theta rhythm supplies a temporal dimension to the processing of auditory information and may be acting as an internal clock as previously suggested (Adey et al., 1966; Pedemonte et al., 1996, 2004; Velluti et al., 2000). The shuffle of the spike series supports the statistic validity of the results because there was no phase locking between the cortical units and theta waves after data 'shuffling'. The phase locking may depend or not on the power of theta hippocampal field potential.

On changing the behavioral state, a temporal relationship (phase locking) was found during wakefulness, SWS, and PS. In addition, this correlation may shift when neurons are acoustically stimulated, and the same neurons could show different correlation for spontaneous and evoked activities. The influence that attention processes exert on hippocampal activity may indicate a point of interaction between those processes and the changes in the pattern of discharge of auditory neurons in sleep and wakefulness. Our results are indicative of a new approach to sensory processing analysis in relation to behavioral states and particularly with all sleep stages.

The sleep-waking dissociation between evoked and spontaneous unitary activity was reported by Huttenlocher (1960) for auditory units recorded from the mesencephalic reticular formation. Examples of a divergent control between spontaneous and evoked firing rates were reported at auditory nuclei as well as at the cortical level, suggesting a differential central control mode (Morales-Cobas et al., 1995; Peña et al., 1999). The phase-locking differences observed between spontaneous and evoked discharges are consistent with the existence of multiple synaptic inputs converging over the neuron. Thus, the observed differences in the hipp theta-unit phase locking of spontaneous and evoked activity, reported by Pedemonte et al. (2001) is also suggestive of a different mode of central control.

We may assume that the same cell may be active during different behaviors although engaged with different networks. Additionally, it is well known that hippocampal place unit activity represents an environmental map associated with behavior and sensory input, as demonstrated by O'Keefe and Recce (1993). Also, a temporal or spatial positioning was reported by Wallestein et al. (1998) for hippocampal units. Both approaches suggest the necessity of a close temporal control of the input signals that could be supplied by the hipp theta rhythm, a low frequency quasisinusoidal rhythm.

The demonstration of a correlation between the incoming information and the hipp theta rhythm opens a new way to study the cortical processing, not only in W periods but also during sleep. The relationship established between neuronal discharges and the hipp theta rhythm may represent a possible link between attention mechanisms, wakefulness/sleep, and auditory processing or, at least, a temporal input organization, supporting the notion of an internal clock role in W and, most intriguing, during sleep (Pedemonte et al., 2001).

Acoustic Communication in Noise

Humans are remarkably adept at acoustic communication in high levels of background noise (the cocktail-party effect). For efficient communication the average speech level should exceed that of the noise by 6 dB. But speech may be intelligible at negative signal-to-noise ratios for continuous speech in which the listener is familiar with the subject matter or if the speech and the noise are separated in space (known as *spatial release from masking*).

Recently, several species of ranid frogs have been shown to produce a wide variety of vocalizations containing significant ultrasonic (>20 kHz) harmonics (Narins and Hurley, 1982). In two genera of Asian frogs (*Odorrana* and *Huia*), auditory evoked potentials and single unit recordings from the inferior colliculus confirm auditory sensitivity up to 39 kHz. It is believed that ultrasonic sensitivity in these animals has evolved in response to selection pressure from the high-level broadband ambient noise produced by the rushing streams in these animals' environments.

Peter Narins
University of California, Los Angeles, USA

COMPLEX SOUND PROCESSING AT THE AUDITORY SINGLE UNIT LEVEL

Some experimental works have been conducted in sleeping animals even with artificial, simple stimulus, while others employed complex sounds, such as the natural animal call.

Different neuronal types, responding to sound, are present in the inferior colliculus and the auditory cortex. For instance, Tammer et al. (2004) reported diverse grouping of neurons related to natural sound, vocalizations, in the squirrel monkey inferior colliculus. Type 1 neurons are activated by self-produced vocalizations as well as vocalizations of group mates and nonspecies-specific sounds. Type 2 neurons are activated by vocalizations of group mates and other acoustic stimulus although not by self-produced calls. Type 3 units are activated by self-produced vocalization but not by group calls or other external sounds (Tammer et al., 2004).

In our experimental paradigm we tested cortical auditory neuron response to artificial sounds, clicks or pure tones at the unit characteristic frequency. Afterward, the unit response to a prerecorded guinea pig call, 700 ms duration, was analyzed. In addition, the natural sound stimulus was delivered in a direct or inverted in time manner, evoking a different single unit response on passing from a waking state to SWS (Fig. 5.15);

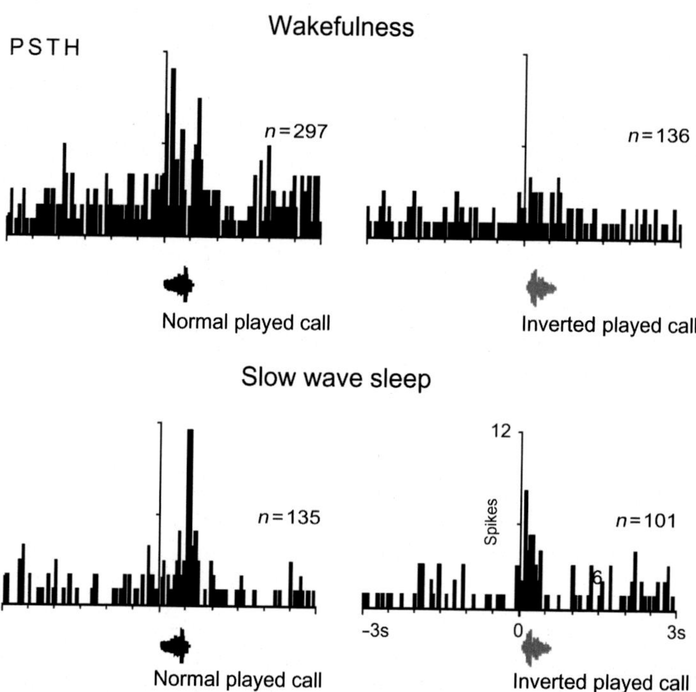

Figure 5.15 Response of an auditory cortex (AI) neuron evoked by a guinea pig prerecorded natural call ('whistle'). It was played back in its natural mode, direct, and reversed in time, during wakefulness and slow wave sleep. The peristimulus time histogram (PSTH) during a wakefulness epoch exhibits the unit firing to the direct natural call decreasing by half when the same sound was played backward, with an associated change in the pattern. During slow wave sleep the unit also exhibited firing shifts, depending on whether the sound was played normally or inverted in time. Thus, the units (included in some neuronal network) may recognize the difference between direct and inverted in time stimuli even in sleep. The response to the stimulus that is inverted in time in slow wave sleep is smaller and has shorter latency. A relation to the large part of the stimulus appears as possible. *Modified from Pérez-Perera, L., Bentancor, C., Pedemonte, M., Velluti, R.A., 2001. Auditory cortex unitary activity correlated to sleep-wakefulness and theta rhythm in response to natural sounds. Actas de Fisiología 7:187; Pérez-Perera, L., 2002. Actividad unitaria de la corteza auditiva: ritmo theta del hipocampo y respuesta a vocalizaciones en el ciclo vigilia-sueño. Master Thesis, Montevideo.*

i.e., these auditory neurons (and their neuronal network) can distinguish the differences even in sleep at the cortical as well as at the inferior colliculus level.

It is not known whether the lack of inferior colliculus unit activity during self-produced vocalization is due to a direct inhibitory input or the result of an inhibition on a second structure normally providing the external nucleus with auditory input. Anatomical studies have shown that the external nucleus is reciprocally connected with the periaqueductal gray (Radmilovich et al., 1991), a structure considered to be an important vocalization control area. The existence of such a direct connection suggests that the periaqueductal vocalization area exerts a gating function on the auditory input reaching the external nucleus of the inferior colliculus. This function might serve to reduce awareness of self-produced vocalizations in relation to, say, conspecific calls and other external stimuli. Neurons that are activated by external sounds but not by self-produced vocalizations have been found also in the auditory cortex. The fact that such neurons exist in the inferior colliculus suggests that selective attention processes already take place at midbrain level (Tammer et al., 2004; Pérez-Perera et al., 2001; Pérez-Perera, 2002; Bentancor et al., 2006).

When guinea pigs were stimulated with their own call, a 'whistle' of 700 ms duration, the response of auditory cortex neurons was different in W and sleep. Furthermore, when the natural call was played backwards, i.e., inverted in time, the neuronal firing changed in wakefulness as well as during SWS. As depicted in Fig. 5.16, the response to a 'whistle' in wakefulness decreased when the same vocalization or call was presented reversed in time.

The same cortical auditory neuron recorded during a SWS epoch exhibited a totally different evoked response. The PSTH generated by normally played stimulus showed a peak in close temporal correlation with the 'whistle' higher amplitude. On the other hand, when the natural call was played inverted in time, the evoked unit firing and latency decreased. Perhaps, the PSTH unit discharge peak was relevant in relation to the highest amplitude of the stimulus; in any case, it is evident that the cortical processing of complex sounds continues to be carried out in SWS.

The reported relationship between the auditory neurons and the hippocampal theta waves, phase locking, was also present when the natural call stimulation was used. The cortical neurons studied exhibited cross-correlation (phase locking) during W, SWS, and PS (Pérez-Perera et al., 2001; Pérez-Perera, 2002).

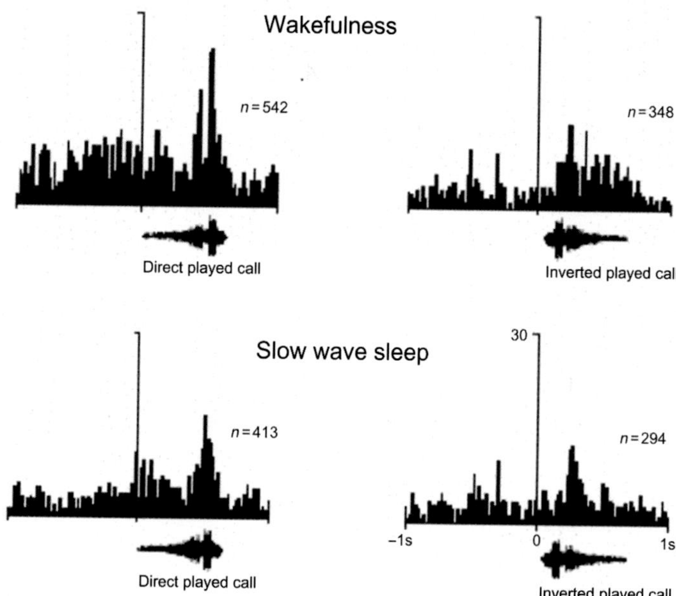

Figure 5.16 Umitary activity modulated by sleep and by stimuli.

Inferior colliculus neurons have also shown a differential response to natural sound stimulation when presented direct or inverted in time (Fig. 5.16). Two factors may be involved in the response changes to direct or inverted sound stimulation: (1) a different temporal organization of the stimuli frequencies is present when presented backward, or (2) it may be also possible that the differences express a new 'significance'. Anyway, the unit and its associated neuronal network can distinguish both stimuli even in SWS (Bentancor et al., 2006).

A similar approach in birds also introduced auditory units firing changes during SWS, stimulating with the bird's own song. The bird song system has neurons that respond to auditory stimuli, particularly to the bird's own song. Recordings at the high vocal center revealed that the single unit response to auditory stimuli is great during SWS monitored by EEG recordings. Spontaneous waking causes the end of this firing within milliseconds. It is likely that if the vocal control system replays the song, 'then the replay functions to maintain the memory of the song in the adult' (Nick and Konishi, 2001).

REFERENCES

Adey, W.R., Dunlop, C.W., Hendrix, C.E., 1960. Hippocampal slow waves distribution and phase relations in the course of approach learning. Arch. Neurol. 3, 74–90.

Adey, W.R., Kado, R.T., McIlwain, J.T., Walter, D.O., 1966. The role of neuronal elements in regional cerebral impedance changes in alerting, orienting and discriminative responses. Exp. Neurol. 15, 490–510.

Affani, J.M., Cervino, C.O., 2005. Interactions between sleep, wakefulness and the olfactory system. In: Parmeggiani, P.L., Velluti, R.A. (Eds.), The Physiologic Nature of Sleep. Imperial College Press, London.

Bentancor, C., Pedemonte, M., Velluti, R.A., 2006. Actividad neuronal del colículo inferior durante el ciclo sueño-vigilia en respuesta a vocalizaciones de la especie. Physiol. Mini-Rev. 2, 65.

Best, P.J., White, A.M., Minai, A., 2001. Spatial processing in the brain: the activity of hippocampal place cells. Ann. Rev. Neurosci. 24, 459–486.

Braun, A.R, Balkin, T.J, Wesensten, N.J., Carson, R.E., Varga, M., Baldwin, P., et al., 1997. Regional cerebral blood flow throughout the sleep-wake cycle. An H2(15)O PET study. Brain 120, 1173–1197.

Braun, A.R., Balkin, T.J., Wesesten, N.J., Gwadry, F., Carson, R.E., Varga, M., et al., 1998. Dissociated pattern of activity in visual cortices and their projections during human rapid eye movement sleep. Science 279, 91–95.

Buño, W., Velluti, J.C., 1977. Relationship of hippocampal theta cycle with bar pressing during self-stimulation. Physiol. Behav. 19, 615–621.

Burgess, A.P., Gruzelier, J.H., 1997. Short duration synchronization of human theta rhythm during recognition memory. Neurol. Rep. 8, 1039–1042.

Buszaki, G., 1996. The hippocampus-neocortical dialogue. Cereb. Cortex 6, 61–92.

Cantero, J.L., Atienza, M., Stickgold, R., Kahana, M.J., Madsen, J.R., Kocsis, B., 2003. Sleep-dependent θ oscillations in the human hippocampus and neocortex. J. Neurosci. 23, 10893–10897.

Cazard, P., Buser, P., 1963. Modification des résponses sensorielles corticales par stimulation de l'hippocampe dorsal chez le lapin. Electroenceph. Clin. Neurophysiol. 15, 413–425.

Doppelmayr, M., Klimesch, W., Schwaiger, J., Auinger, P., Winkler, T., 1998. Theta synchronization in the human EEG and episodic retrieval. Neurosci. Lett. 257, 41–44.

Edeline, J.-M., 2003. The thalamo-cortical auditory receptive fields: regulation by the sates of vigilance, learning and neuromodulatory systems. Exp. Brain Res. 153, 554–572.

Edeline, J.-M., 2005. Learning-induced plasticity in the thalamo-cortical auditory system: should we move from rate to temporal code descriptions? In: Konig, R., Heil, P., Budinger, E., Scheich, H. (Eds.), The Auditory Cortex. Lawrence Erlbaun Ass, Mahwah, New Jersey, London, pp. 365–382.

Edeline, J.-M., Manunta, Y., Hennevin, E., 2000. Auditory thalamus neurons during sleep: changes in frequency selectivity, threshold, and receptive field size. J. Neurophysiol. 84, 934–952.

Gambini, J.P., Velluti, R.A., Pedemonte, M., 2002. Hippocampal theta rhythm synchronized visual neurones in sleep and waking. Brain Res. 926, 137–141.

García-Austt, E., 1984. Hippocampal level of neural integration. In: Ajmone-Marsan, E., Reinoso-Suárez, F. (Eds.), Cortical Integration Basic Archicortical and Cortical Association Levels of Neuronal Integrations. IBRO Monograph Series, Raven Press, New York, pp. 91–104.

Garzón, M., 1996. Estudio morfofuncional de los núcleos reticular oral y reticular caudal del tegmento pontino como regiones generadoras de sueño paradójico. Tesis Doctoral, Universidad Autónoma de Madrid.

Gaztelu, J.M., Romero-Vives, M., Abraira, V., García-Austt, E., 1994. Hippocampal EEG theta power density is similar during slow-wave sleep and paradoxical sleep. A long-term study in rats. Neurosci. Lett. 172, 31–34.

Grastyán, E., Lissák, K., Madarász, I., 1959. Hippocampal activity during the development of conditioned reflex. Electroenceph. Clin. Neurophysiol. 11, 409–430.

Huttenlocher, P.R., 1960. Effects of the state of arousal on click responses in the mesencephalic reticular formation. Electroencephgr. Clin. Neurophysiol. 12, 819–827.

Issa, E.B., Wang, X., 2008. Sensory responses during sleep in primate primary and secondary auditory cortex. J. Neurosci. 28 (53), 14467–14480.

Kahana, M.J., Sekuler, R., Caplan, J.B., Kirschen, M., Madsen, J.R., 1999. Human theta oscillations exhibit task dependence during virtual maze navigation. Nature 399, 781–784.

Kahana, M.J., Seelig, D., Madsen, J.R., 2001. Theta returns. Curr. Opinion Neurobiol. 11, 739–744.

Klimesch, W., 1999. EEG alpha and theta oscillations reflect cognitive and memory performance: a review and analysis. Brain Res. Rev. 29, 169–195.

Kocsis, B., Vertes, R.P., 1992. Dorsal raphe neurones: synchronous discharge with theta rhythm of the hippocampus in the freely behaving rat. J. Neurophysiol. 68, 1463–1467.

Komisariuk, B., 1970. Synchrony between limbic system theta activity and rhythmical behaviour in rats. J. Comp. Physiol. Psychol. 10, 482–492.

Kramis, R., Vanderwolf, C.H., Bland, B.H., 1975. Two types of hippocampal rhythmical slow activity in both the rabbit and the rat: relations to behaviour and effects atropine, diethyl ether, urethane and pentobarbital. Exp. Neurol. 49, 58–85.

Lerma, J., García-Austt, E., 1985. Hippocampal theta rhythm during paradoxical sleep. effects of afferent stimuli and phase-relationships with phasic events. Electroenceph. Clin. Neurophysiol. 60, 46–54.

Liberman, T., Velluti, R.A., Pedemonte, M., 2006. Correlación entre neuronas auditivas del colículo inferior y el ritmo theta del hipocampo. Physiol. Mini-Rev. 2, 65.

Lopez de Silva, F.H., Witter, M.P., Boeijinga, P.H., Lohman, A.H.M., 1990. Anatomical organization and physiology of the limbic cortex. Physiol. Rev. 70, 453–511.

Magee, J.C., 2003. A prominent role for intrinsic neuroneal properties in temporal coding. Trends Neurosci. 26, 14–16.

Margrie, T.W., Schaefer, A.T., 2003. Theta oscillation coupled spike latencies yield computational vigour in a mammalian sensory system. J. Physiol 546, 363–374.

McNaughton, N., Morris, R.G., 1987. Chlordiazepoxide, an anxiolytic benzodiazepine, impairs place navigation in rats. Behav. Brain Res. 24, 39–46.

Metha, M.R., Lee, A.K., Wilson, M.A., 2002. Role of experience and oscillations in transforming a rate code into a temporal code. Nature 417, 741–746.

Morales-Cobas, G., Ferreira, M.I., Velluti, R.A., 1995. Sleep and waking firing of inferior colliculus neurons in response to low frequency sound stimulation. J. Sleep Res. 4, 242–251.

Narins, P.M., Hurley, D.D., 1982. The relationship between call intensity and function in the Puerto Rican Coqui (Anura: Leptodactylidae). Herpetologica 38, 287–295.

Nick, T.A., Konishi, M., 2001. Dynamic control of auditory activity during sleep: correlation between song response and EEG. PNAS 98, 14012–14016.

Nir, Y., 2016. Responses in rat core auditory cortex are preserved during sleep spindle oscillations. Sleep. 39 (5), 1069–1082.

Nuñez, A., de Andrés, I., García-Austt, E., 1991. Relationships of nucleus reticularis pontis oralis neuronal discharge with sensory and carbachol evoked hippocampal theta rhythm. Exp. Brain Res. 87, 303–308.

O'Keefe, J., Recce, M.L., 1993. Phase relationship between hippocampal place units and EEG theta rhythm. Hippocampus 3, 317–330.

Parmeggiani, P.L., Rapisarda, C., 1969. Hippocampal output and sensory mechanisms. Brain Res. 14, 387–400.

Parmeggiani, P.L., Lenzi, P., Azzaroni, A., D'Alessandro, R., 1982. Hippocampal influence on unit responses elicited in the Cat'S auditory cortex by acoustic stimulation. Exp. Neurol. 78, 259–274.

Pedemonte, M., Velluti, R.A., 2005. What individual neurons tell us about encoding and sensory processing in sleep. In: Parmeggiani, P.L., Velluti, R.A. (Eds.), The Physiologic Nature of Sleep. Imperial College Press, London, 489–408.

Pedemonte, M., Peña, J.L., Morales-Cobas, G., Velluti, R.A., 1994. Effects of sleep on the responses of single cells in the lateral superior olive. Arch. Ital. Biol. 132, 165–178.

Pedemonte, M., Peña, J.L., Velluti, R.A., 1996. Firing of inferior colliculus auditory neurone is phase-locked to the hippocampus theta rhythm during paradoxical sleep and waking. Exp. Brain Res. 112, 41–46.

Pedemonte, M., Rodríguez, A., Velluti, R.A., 1999. Hippocampal theta waves as an electrocardiogram rhythm timer in paradoxical sleep. Neurosci. Lett. 276, 5–8.

Pedemonte, M., Perez-Perera, L., Peña, J.L., Velluti, R.A., 2001. Sleep and wakefulness auditory processing: cortical units vs hippocampal theta rhythm. Sleep Res. Online 4, 51–57.

Pedemonte, M., Goldstein-Daruech, N., Velluti, R.A., 2003. Temporal correlation between heart rate, medullary units and hippocampal theta rhythm in anesthetized, sleeping and awake guinea pigs. Auto. Neuroc. Basic Clin. 107, 99–104.

Pedemonte, M., Drexler, D.G., Velluti, R.A., 2004. Cochlear microphonic changes after noise exposure and gentamicin administration during sleep and waking. Hear. Res. 194, 25–30.

Pedemonte, M., Gambini, J.P., Velluti, R.A., 2005. Novelty-induced correlation between visual neurons and the hippocampal theta rhythm in sleep and wakefulness. Brain Res. 1062, 9–15.

Peña, J.L., Pedemonte, M., Ribeiro, M.F., Velluti, R., 1992. Single unit activity in the guinea-pig cochlear nucleus during sleep and wakefulness. Arch. Ital. Biol. 130, 179–189.

Peña, J.L., Pérez-Perera, L., Bouvier, M., Velluti, R.A., 1999. Sleep and wakefulness modulation of the neuronal fi ring in the auditory cortex of the guinea-pig. Brain Res. 816, 463–470.

Pérez-Perera, L., 2002. Actividad unitaria de la corteza auditiva: ritmo theta del hipocampo y respuesta a vocalizaciones en el ciclo vigilia-sueño. Master Thesis, Montevideo.

Pérez-Perera, L., Bentancor, C., Pedemonte, M., Velluti, R.A., 2001. Auditory cortex unitary activity correlated to sleep-wakefulness and theta rhythm in response to natural sounds. Actas de Fisiología 7, 187.

Pompeiano, O., 1970. Mechanisms of sensorimotor integration during sleep. In: Stellar, E., Sprague, JM. (Eds.), Progress in Physiological Psychology. Academic Press, New York/London, pp. 1–179.

Radmilovich, M., Bertolotto, C., Peña, J.L., Pedemonte, M., Velluti, R.A., 1991. A search for e mesencephalic pariaqueductal gray-cochlear nucleus connection. Acta Physiol. Pharmacol. Latinoam. 41, 369–375.

Rawlins, J.N.P., 1985. Associatios across time in the hippocampus as a temporary memori store. Behav. Brain Sci. 8, 479–496.

Redding, F.K., 1967. Modification of sensory cortical evoked potentials by hippocampal stimulation. Electroenceph. Clin. Neurophysiol. 22, 74–83.

Reinoso-Suárez, F., De Andrés, I., Rodrigo-Angulo, M.L., Rodríguez-Veiga, E., 1994. Location and anatomical connections of a paradoxical sleep induction site in the cat ventral pontine tegmentum. Eur. J. Neurosci. 6, 1829–1836.

Reivich, M., 1974. Blood flow metabolism couple in brain. In: Plum, F. (Ed.), Brain Dysfunction in Metabolic Disorders. Raven, New York, pp. 125–140.

Saldaña, E., Feliciano, M., Mugnaini, E., 1996. Distribution of descending projections from primary auditory neocortex to inferior colliculus mimics the topography of intracollicular projections. J. Comp. Neurol. 371, 15–40.

Sela, Y., Vyazovskiy, V.V., Cirelli, C., Tononi, G., Nir, Y., 2016. Responses in rat core auditory cortex are preserved during sleep spindle oscillations. Sleep. 39 (5), 1069–1082.

Skaggs, W.E., McNaughton, B.L., Wilson, M.A., Barnes, C., 1996. Theta phase precession in hippocampal neuroneal populations and the compression of temporal sequences. Hippocampus 6, 149–172.

Tammer, R., Ehrenreich, L., Jürgens, U., 2004. Telemetrically recorded neuronal activity in the inferior colliculus and bordering tegmentum during vocal communication in squirrel monkeys (*Saimiri sciureus*). Behav. Brain Res. 151, 331–336.

Velluti, R.A., 1997. Interactions between sleep and sensory physiology. A review. J. Sleep Res. 6, 61–77.

Velluti, R.A., 2005. Remarks on sensory neurophysiological mechanisms participating in active sleep processes. In: Parmeggiani, P.L., Velluti, R.A. (Eds.), The Physiologic Nature of Sleep. Imperial College Press, London, pp. 247–265.

Velluti, R.A., Pedemonte, M., 2002. *In Vivo* approach to the cellular mechanisms for sensory processing in sleep and wakefulness. Cell Mol. Neurobiol. 22, 501–516.

Velluti, R.A., Pedemonte, M., 2012. Sensory neurophysiologic functions participating in active sleep processes. Sleep Sci. 5 (4), 103–106.

Velluti, R.A., Pedemonte, M., García-Austt, E., 1989. Correlative changes of auditory nerve and microphonic potentials throughout sleep. Hearing Res. 39, 203–208.

Velluti, R.A., Pedemonte, M., Peña, J.L., 1990. Auditory brain stem unit activity during sleep phases. In: Horne, J. (Ed.), Sleep '90. Pontenagel Press, Bochum, pp. 94–96.

Velluti, R.A., Pedemonte, M., Peña, J.L., 2000. Reciprocal actions between sensory signals and sleep. Biol. Signals Recept. 9, 297–308.

Vertes, R.P., Kocsis, B., 1997. Brainstem–Diencephalo–Septohippocampal systems controlling the theta rhythm of the hippocampus. Neuroscience 81, 893–926.

Vinogradova, O.S., 2001. Hippocampus as comparator: role of the two input and two output systems of the hippocampus in selection and registration of information. Hippocampus 11, 578–598.

Wallestein, G.W., Eichenbaum, H., Hasselmo, M.E., 1998. The hippocampus as an associator of discontiguous events. Trends Neurosci. 21, 317–323.

Winson, J., 1978. Loss of hippocampal theta rhythm results in spatial memory deficit in the rat. Science 201, 160–163.

FURTHER READING

Naka, D., Kakigi, R., Hoshiyama, M., Yamasaki, H., Okusa, T., Koyama, S., 1999. Structure of the auditory evoked magnetic fields during sleep. Neuroscience 93, 573–583.

Soja, P.J., Cairns, B.E., Kristensen, M.P., 1998. Transmission through ascending trigeminal and lumbar sensory pathways: dependence on behavioral state. In: Lydic, R., Baghdoyan, H.A. (Eds.), Handbook of Behavioral State Control. CRC Press, Boca Raton/London/New York/Washington, pp. 521–544.

CHAPTER 6

Auditory Influences on Sleep

. . .***Sleep appears to be a behavioural state resulting from dynamic interactions of different physiologic functions*** *in response to several endogenous (feeding, fatigue, temperature, instinctive drives) and exogenous (light-dark, temperature, food, season, social drives) cues. From the viewpoint of its determination, the mechanism appears complex as to justify a theoretical distinction between* ***proximate***, ***intermediate*** *and* ***remote*** *aspects of determination of sleep behaviour. This gradual approach to sleep behaviour avoids extending the category of rigid causal determination beyond the molecular and cellular levels and forcing experimental results to fit a reductionistic theory in spite of the fact that many elementary physiologic events characterising sleep behaviour are not specific to sleep alone. In other words, sleep, like wakefulness, is a function of other interactive functions and not the unique result of the compelling influence of a segregated and highly specific neuronal network of the central nervous system.*

– ***Prof. P.L. Parmeggiani, 2008, Universitá di Bologna, Italy***

The constantly present sensory incoming signals imposing conditions on the brain activity and, vice versa, the state of the brain imposing rules to the incoming information, is now the main objective (Fig. 6.1). The efferent pathways are the channels connecting the central nervous system (CNS) with receptors and nuclei, such as the auditory efferent system arriving at the cochlea and nuclei (Chapter 1: Brief Analysis of the Auditory System Organization and Its Physiologic Basis). Furthermore, in the visual system at sleep also modifies the rat electroretinogram (Galambos et al., 1994) and exerts actions on ganglion cells (Cervetto et al., 1976).

In spite of the many membrane or circuital oscillators described, my view stresses the capacity of the sensory incoming information, from the outer and the inner (body) worlds, acting as a partial although relevant generator of the CNS basic activity, which performs continuously during wakefulness as well as during sleep. It is my view that the sensory input is a decisive physiological fact that, acting on different brain areas, may support the basic brain activity during sleep as well. Wakefulness or sleep characteristics have to be modulated based on this continuous incoming flow of information (Fig. 6.2).

The Auditory System in Sleep
DOI: https://doi.org/10.1016/B978-0-12-810476-7.00006-3

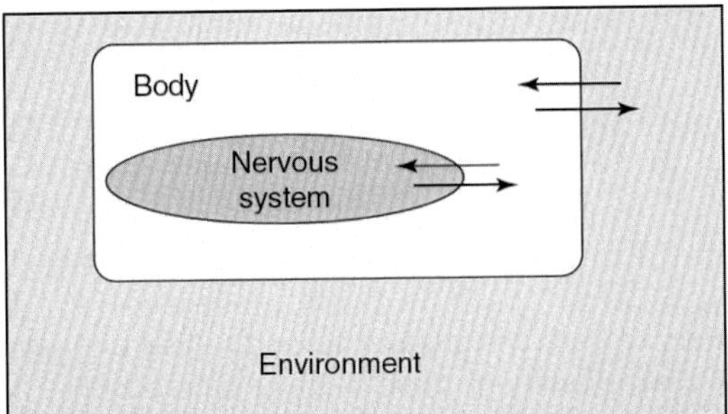

Figure 6.1 The nervous system placed inside the body has communicating channels that connect it with the environment and with the body itself through a diverse set of receptors.

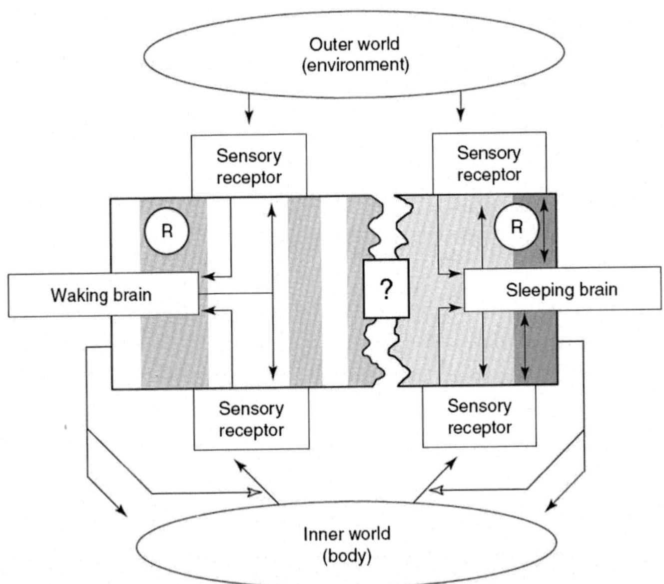

Figure 6.2 The waking and sleeping brain receives, during its different phases, sensory information from the outer world and the inner (body) world; this means that an enormous and continuous barrage of incoming activity impinges on the brain, activating or deactivating diverse regions of the CNS. All the receptors are controlled by efferent connections and CNS actions are exerted on the body's physiology during both wakefulness and sleep. The inner world receptors carry information from viscera, muscles, tendons, and joints, from pressure and chemoreceptors, blood flowing noise, and so forth. There are also intracerebral receptors for blood sugar levels, hormones, and the like.

When some still unknown signals cause the system to change the neural networks in order to enter the "sleeping mode," the sensory systems change along with it, shifting its wakefulness control characteristics into a "sensory sleeping mode" function, thus producing a different processing form, we have to admit, according to sleep sensory-related needs:

- initiate and maintain sleep,
- keep a relative isolation from the environment,
- monitor the environment,
- produce wakefulness if the incoming information is significant or of high intensity,
- cooperate with the dream process or content,
- process memory-stored data,
- store new or processed data in memory.

SENSORY INPUT AND SLEEP

The analysis of sensory functions during the sleep-waking cycle leads to the conclusion that normal sleep depends in many ways on the sensory input. It is suggested that the sleep and waking control networks are modulated by several inputs, and therefore a proportion of "passive" effects must be associated with active functions for entering and maintaining normal sleep.

In general, the sensory input is a relevant signal. The total amount of sleep increments is produced under some experimental conditions facing sensory shifts (Velluti, 1997):

- Continuous somatosensory stimulation induces electroencephalography (EEG) synchronization and sleep.
- Total darkness increases sleep although just during a few days.
- Total silence, after bilateral cochlear destruction, increases the amount of sleep (Pedemonte et al., 1996).
- Use of intracochlear implants in deaf humans shows sleep stages percentages to be different when compared with same patients before recovering hearing (Velluti et al., 2003, 2010).
- Lack of olfactory input after epithelial receptor lesions evokes a different proportion (%) of sleep stages and waking (López et al., 2007).

Partial increments in the occurrence of specific sleep stages are observed (review, Velluti, 1997):

- when rats are stimulated with sounds during any sleep stage,

- during stimulation with bright light, which produces slow wave sleep increases in humans,
- during electrical stimulation of the olfactory bulb, which increases slow wave sleep in cats.

The sensory influences on sleep, such as the abolition or decrement of a sleep sign or stage, are produced by

- Continuous light stimulation in rats that decreases paradoxical sleep up to 20 days.
- Bilateral lesions of some vestibular nuclei that abolish rapid eye movements during paradoxical sleep up to 36 days.
- A long exposure to cold that produces decrement of paradoxical sleep to its deprivation.
- Olfactory bulbectomy decreases paradoxical sleep episode incidence and its total amount for up to 15 days.

The lack of sensory inputs as well as their enhancement can produce sleep/waking imbalances, augmenting or diminishing sleep stage proportions and provoking electrophysiological shifts. Thus, the induced changes in the waking and sleep networks lead to imbalances not simply explained by passive sleep production but also by introducing sensory sleep-active influences. A diversity of approaches supports this notion:

1. Sleep and sound are closely related. Environmental noise, as well as regular, monotonous auditory stimuli, such as a mother's lullaby, influence impeding sleep or facilitate it.
2. The CNS and auditory system bioelectrical field activity-evoked potentials, magnetoencephalographic (MEG) evoked activity, shown from early electrophysiological studies; vary in close correlation with wakefulness epochs and especially during sleep stages.
3. The auditory system single neuronal firing exhibits a variety of changes along the different nuclei and primary cortical loci linked to the sleep-wakefulness cycle in many ways: (a) increasing or decreasing firing on passing into sleep, (b) firing during sleep as during wakefulness, (c) changing the discharge pattern, (d) exhibiting theta rhythm phase locking, with no auditory neuron stopping its firing on passing to sleep. Furthermore, Edeline (2003, 2005) and Edeline et al. (2001) reported changes in the receptive field of cortical auditory neurons.

Therefore, it can be concluded that, when asleep, many auditory units are active, probably in association with diverse sleep relevant cell assemblies. Moreover, when functionally shifting into a different neuronal network or cell assembly, a unit may contribute to the sleep process just by increasing, decreasing, or showing no firing shift, according to the new role in the new cell assembly association (Velluti, 2005; Velluti and Pedemonte, 2012).

4. A MEG approach described anatomical place shifts of the sound evoked dipole in the human primary auditory cortex (*planum temporale*) on passing from wakefulness into sleep stage II (Naka et al., 1999; Kakigi et al., 2003). The dipole anatomical position shift obtained with MEG in the *planum temporale* is indicative of a change to a new neuronal group, a different neuronal network, already supported by single unit studies (see Chapter 5: Auditory Unit Activity in Sleep). The MEG auditory evoked activity during sleep (its dipole) appears in a different cortical region from that during wakefulness, thus suggesting a new cell assembly/neuronal network participation.
5. The functional magnetic resonance imaging (fMRI), when combined with EEG recording, showed that the auditory stimuli produce bilateral activation in the human auditory cortex, parietal and frontal areas, both during wakefulness and sleep (Portas et al., 2000).
6. Positron emission tomography exhibits the cerebral blood flow (CBF) measurements in waking and sleep stages (Maquet et al., 2005). It is assumed that shifts in neuronal firing (glia and neuron activity) and oxygen consumption by the tissue induce proportional changes in local CBF, flow-activity coupling. An example of such possibility is the oxygen availability recording using an oxygen cathode (Velluti et al., 1965). When recording from a cat visual cortex the oxygen availability rises when flashes of light are presented to the animal (Fig. 6.3). The lower record shows the oxygen cathode response capacity to a changing air O_2 concentration, that is, when the animal breathes 100% oxygen.

The data exhibited by fMRI strongly support the notion that the sleeping brain is able to process information, detecting meaningful events, as it can be observed in the unitary response in guinea pigs when a complex stimulus (the animal call) is played normally or in reverse (Chapter 5: Auditory Unit Activity in Sleep). A strong interaction between auditory input and sleep processes is put forward.

Some sleep researchers are, unconsciously, looking for a "sleep centre" that does not exist. A CNS centre may be real and useful for controlling functions such as the cardiovascular and the respiratory, while on the other hand, sleep is not a function but a complete different CNS state. This means different brains for the diverse wakefulness conditions, for

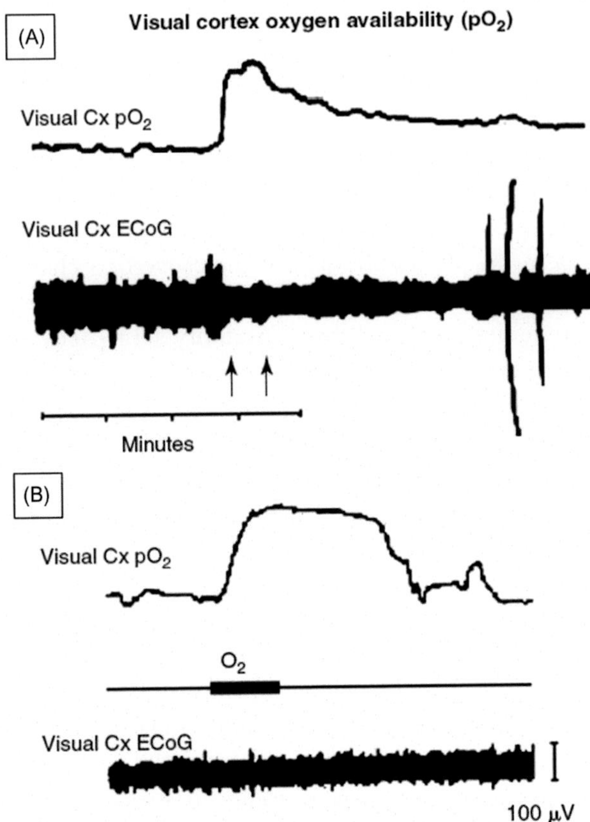

Figure 6.3 Cortical oxygen availability (pO_2) during visual sensory stimulation in an awake cat. In (A), an increase in the oxygen availability appears after light flashes (20/s) are presented for a few seconds to the animal's eyes with dilated pupils. In (B), it is shown the oxygen cathode response capacity when the animal breathes pure oxygen. This technique does not permit calibrating the cathode because the diffusion coefficient at the electrode tip is never known.

sleep stages I, II, slow wave, and for paradoxical sleep with or without phasic components. Hence, sleep means a whole change of networks/cell assemblies, a new cooperative interaction among them, considering that a single network may subserve several different functions. The auditory dipole position change on passing into sleep appears as an advance, experimental evidence, supporting the notion of a new configuration of neuronal networks all over the brain appearing when asleep.

A simple mollusc brain may serve as an example of functionally changing networks associated with different behaviours. The imbalance introduced by cell 1 activity in the *Tritonia* uncomplicated brain produces two functional networks related to two different movements (see Fig. 3.3 of Chapter 3: Notes on Information Processing).

In a complex brain such imbalance may be carried out by a summation of physiological factors that, after some unknown signal or signals, perhaps decreasing light intensity, produce sleep; that is, a group of neuronal networks/cell assemblies progressively begin to prepare the brain into enter sleep. Partially supporting this assumption is the observation that, when a human or animal is passing into sleep, the many variables usually recorded never occur in synchrony but appear with seconds of difference among them, for example, EEG slow activity, electromyogram decrement, eye movements, hippocampal theta rhythm frequency and amplitude, heart rate shifts, arterial pressure changes, breathing rhythm alterations, and so on.

HUMAN AND ANIMAL EXPERIMENTAL DATA

Effects of Sound Stimulation and Auditory Deprivation on Sleep

The organization of human sleep is extremely sensitive to acoustic stimuli (Croome, 1977), and noise generally exerts an arousing influence on it (Muzet and Naitoh, 1977). A noisy nighttime ambiance leads to a decrease in total sleep time, particularly that of delta wave sleep (Stage IV) and paradoxical sleep, with the consequent increase in the time spent in Stage II and wakefulness (Vallet and Mouret, 1984; Terzano et al., 1990). Moreover, the remarkable sleep improvement after noise abatement (Vallet, 1982), suggests that the environment is being continuously scanned by the auditory system. This notion is also supported by results from single unit recordings of sleeping animals previously described (Chapter 5: Auditory Unit Activity in Sleep).

Conflicting results have been reported concerning sound stimulation effects on sleep in animals. Using continuous high intensity white-noise stimulation resulted in almost complete deprivation of paradoxical sleep in rabbits (Khazan and Sawyer, 1963), while in rats the same stimulation led to paradoxical sleep reduction without a decrease in the amount of slow wave sleep (van Twyver et al., 1966). Upon intense auditory stimulation specifically during paradoxical sleep, the number of episodes increased without changing its total amount, while the number of ponto-geniculate-occipital waves was enhanced in cats also stimulated with high intensity sound during the same sleep stage (Ball et al., 1989). Moreover, auditory stimulation carried out more recently in rats led to the conclusion that the pattern of paradoxical sleep occurrence was affected by stimulation both during paradoxical and slow wave sleep, while the total amount of paradoxical sleep was substantially preserved (Amici et al., 2000, 2001). An animal study carried out in rats reported that the SWS-2, with high delta wave power, showed the highest arousal threshold when using a nonmeaningful sound (Neckelmann and Ursin, 1993).

Auditory stimulation in animals as well as in humans seems to produce general actions on sleep processes, still not clearly defined, because stimulation during both sleep stages results in similar changes in the general sleep pattern.

Very interesting results, although old, reported the responsiveness of a conditioned cat to a 5 kHz "positive" tone (reinforced with classical and instrumental conditioning) during wakefulness (Buendía et al., 1963). It was orderly in wakefulness but different in each sleep phase, measured by the tone capacity to awaken the animal. During slow wave sleep the effectiveness of the 5 kHz tone for awakening the animal was not significantly changed, while the surrounding tones ("negatives," never reinforced) capacity decreased. The tone separation ability of the system became sharper, indicative of a more precise discriminating mechanism during slow wave sleep; during paradoxical sleep the cats were practically nonresponsive.

A question arises immediately: Is there any relationship between this discriminative ability during slow wave sleep and the enhanced amplitude of the auditory nerve compound action potential and the increased amplitude of the auditory evoked potentials in all the brain loci studied up to the cortex itself, during this sleep phase (Chapter 1: Brief Analysis of the Auditory System Organization and Its Physiologic Basis)? Moreover, is there any relationship with the increasing auditory single unit firing in slow wave sleep (Chapter 5: Auditory Unit Activity in Sleep)?

On the other hand, total auditory deprivation in guinea pigs by surgical removal of both cochleae enhances slow wave and paradoxical sleep by in a similar proportion while reducing wakefulness, for up to 45 days postlesion (Pedemonte et al., 1996). The slow wave and paradoxical sleep increments cited were determined mainly by an increase in the number of episodes with no change in single episode duration. The authors contend that the relative isolation from the outside world may elicit part of the change observed in deaf guinea pigs. Thus, eliminating an input to a complex set of networks, such as the ones that regulate the sleep-waking cycle, would introduce functional shifts, particularly if such input has some significance, as appears to be the case for the behaviour under study: wakefulness and sleep.

Contrasting results have been reported although all of them are indicative of a correlation between both auditory input and the sleep-waking processes, supported by stimulation experiments as well as by the opposite approach, recording from deaf guinea pigs. More experimental data are clearly needed in the particular field of auditory sensory processes and sleep.

The many technical approaches reviewed support the notion that the sensory information in general and the auditory incoming information in particular exert influences on sleep through a dynamic neuronal participation in different sleep-related cell assemblies. Fig. 6.4, section 7, shows the sleep pattern of a deaf patient that has an intracochlear implant. This patient could be studied as a deaf person or as an almost normal-hearing person. When the sleep was analysed during two nights at the sleep laboratory as deaf patient (intracochlear implant off) the sleep architecture and stages percentages were similar to the normal controls of a similar age. When performing the two-night recording as an almost normal-hearing person (implant on) the results were totally different, with huge changes in the stages percentages. This result shows quite clearly the auditory influence on sleep already observed in deaf guinea pigs (Velluti et al., 2010).

Finally, I previously postulated that the auditory neurons firing in sleep at the same rate and pattern as during wakefulness are those neurons that monitor the environment. These cell types increase their percentages from the brain stem up to the auditory primary cortical level. At the brain stem the units that shift their firing percentages perhaps are participating more closely in sleep-related regions. In the end, the units that increase or decrease their firing are postulated to be sleep-related active neurons, at the cortical as well as the brain stem levels.

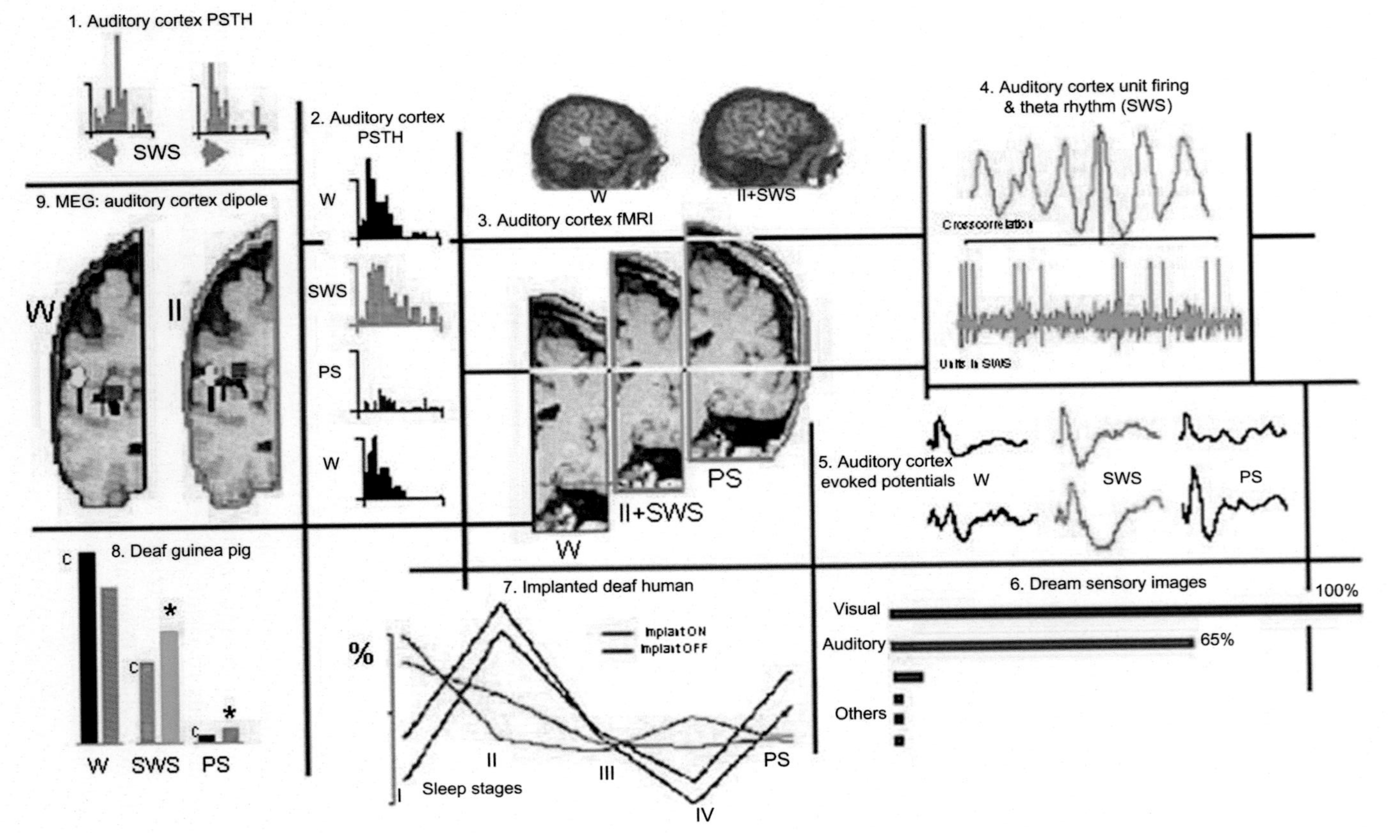

1. Auditory cortex PSTH
SWS
2. Auditory cortex PSTH
W
SWS
PS
W
W
II+SWS
3. Auditory cortex fMRI
W
II+SWS
PS
4. Auditory cortex unit firing & theta rhythm (SWS)
Crosscorrelation
Units in SWS
5. Auditory cortex evoked potentials
W
SWS
PS
6. Dream sensory images
Visual
Auditory
Others
100%
65%
7. Implanted deaf human
%
Implant ON
Implant OFF
I
II
III
IV
PS
Sleep stages
8. Deaf guinea pig
C
*
W
SWS
PS
9. MEG: auditory cortex dipole
W
II

◀ **Figure 6.4** Diverse technical approaches supporting the postulated notion of the importance and possible active participation of the auditory input on sleep processes. Three human half brain tomographic cuts (centre) represent the three main functional possibilities: wakefulness (W), slow wave sleep (SWS), and paradoxical sleep (PS). (1) Poststimulus time histogram (PSTH) changes of a cortical auditory neuron firing shift when stimulated with natural sound played directly or backwards (Pérez-Perera et al. 2001). (2) PSTH of a cortical unit on passing from W to SWS and PS exhibits firing and pattern shifts (Peña et al., 1999). (3) Human auditory cortical imaging (fMRI) demonstrates activity during sleep (modified from Portas et al., 2000). (4) The cortical auditory neurons can be phase locked to hippocampal theta rhythm (Pedemonte et al., 2001). (5) Rat auditory cortical evoked potentials through the sleep-waking cycle show amplitude changes (Hall and Borbély, 1970). (6) The dream auditory "images" are present in 65% of dream recalls (McCarley and Hoffman, 1981). Human and guinea pig deafness influence sleep: (7) the human recorded with the intracochlear implant off and on shows different sleep stages percentages while (8) the guinea pig exhibits (bars) an increase in sleep time with decreasing wakefulness (Velluti et al., 2003; Pedemonte et al., 1996). (9) The human MEG shows a place shift of the dipole evoked by three sound stimulating frequencies on passing into sleep stage II, demonstrating a change of neuronal network/cell assembly. *Modified from Kakigi, R. Naka, D., Okusa, T., Wang, X., Inui, K., Qiu, Y., et al., 2003. Sensory perception during sleep in humans: a magnetoencephalograhic study. Sleep Med. 4, 493–507 and Naka, D., Kakigi, R., Hoshiyama, M., Yamasaki, H., Okusa, T., Koyama, S., 1999. Structure of the auditory evoked magnetic fields during sleep. Neuroscience 93, 573–583.*

REFERENCES

Amici, R., Domeniconi, R., Jones, C.A., Morales-Cobas, G., Perez, E., Tavernese, L., et al., 2000. Changes in REM sleep occurrence due to rhythmical auditory stimulation in the rat. Brain Res. 868, 241–250.

Amici, R., Morales-Cobas, G., Jones, C.A., Perez, E., Torterolo, P., Zamboni, G., et al., 2001. REM sleep enhancement due to rhythmical auditory stimulation in the rat. Behav. Brain Res. 123, 155–163.

Ball, W.A., Morrison, A.R., Ross, R.J., 1989. The effects of tones on PGO waves in slow wave sleep and paradoxical sleep. Exp. Neurol. 104, 251–256.

Buendía, N., Sierra, G., Goode, M., Segundo, J.P., 1963. Conditioned and discriminatory responses in wakeful and sleeping cats. Electroenceph. Clin. Neurophysiol. 24 (Suppl.), 199.

Cervetto, L., Marchiafava, P.L., Pasino, E., 1976. Influence of efferent retinal fibers on responsiveness of ganglion cell to light. Nature 269, 56–57.

Croome, D.J., 1977. Noise and sleep. Noise, Building and People. Pergamon Press, London, pp. 101–109.

Edeline, J.-M., 2003. The thalamo-cortical auditory receptive fields: regulation by the sates of vigilance, learning and neuromodulatory systems. Exp. Brain Res. 153, 554–572.

Edeline, J.-M., 2005. Learning-induced plasticity in the thalamo-cortical auditory system: should we move from rate to temporal code descriptions? In: Konig, R., Heil, P., Budinger, E., Scheich, H. (Eds.), The Auditory Cortex. Lawrence Erlbaun Associates, Mahwah, NJ, London, pp. 365–382.

Edeline, J.M., Dutrieux, G., Manunta, G., Hennevin, E., 2001. Diversity of receptive field changes in auditory cortex during natural sleep. Eur. J. Neurosci. 14, 1865–1880.

Galambos, R., Juhász, G., Kékesi, A.K., Nyitrai, G., Szilágy, N., 1994. Natural sleep modifies the rat electroretinogram. Proc. Natl Acad. Sci. USA 91, 5153–5157.

Hall, R.D., Borbély, A.A., 1970. Acoustically evoked potentials in the rat during sleep and waking. Exp. Brain Res 11, 93–110.

Kakigi, R., Naka, D., Okusa, T., Wang, X., Inui, K., Qiu, Y., et al., 2003. Sensory perception during sleep in humans: a magnetoencephalograhic study. Sleep Med. 4, 493–507.

Khazan, N., Sawyer, C.H., 1963. "Rebound" recovery from deprivation of paradoxical sleep in the rabbit. Proc. Soc. Exp. Biol. Med. 114, 536–539.

López, C., Rodríguez-Servetti, Z., Velluti, R.A., Pedemonte, M., 2007. Influence of the Olfactory System on the Wakefulness-Sleep Cycle. World Federation of Sleep Medicine Sleep Research Societies, Cairns, Australia. Sleep and Biological Rhythms 5, 54.

Maquet, P.A.A., Sterpenich, V., Albouy, G., Dang-vu, T., Desseilles, M., Boly, M., et al., 2005. Brain imaging on passing to sleep. In: Parmeggiani, P.L., Velluti, R.A. (Eds.), The Physiologic Nature of Sleep. Imperial College Press, London, pp. 123–137.

McCarley, R.W., Hoffman, E.A., 1981. REM sleep dreams and the activation-synthesis hypothesis. Am. J. Psychiat 38, 904–912.

Muzet, A., Naitoh, P., 1977. Sommeil et bruit. Confront. Psychiat. 15, 215–235.

Naka, D., Kakigi, R., Hoshiyama, M., Yamasaki, H., Okusa, T., Koyama, S., 1999. Structure of the auditory evoked magnetic fields during sleep. Neuroscience 93, 573–583.

Neckelmann, D., Ursin, R., 1993. Sleep stages and EEG power spectrum in relation to acoustical stimulus arousal threshold in the rat. Sleep 16, 467–477.

Pedemonte, M., Peña, J.L., Torterolo, P., Velluti, R.A., 1996. Auditory deprivation modifies sleep in the guinea-pig. Neurosci. Lett. 223, 1–4.

Pedemonte, M., Pérez-Perera, L., Peña, J.L., Velluti, R.A., 2001. Sleep and wakefulness auditory processing: cortical units vs. hippocampal theta rhythm. Sleep Res. Online 4, 51–57.

Peña, J.L., Pérez-Perera, L., Bouvier, M., Velluti, R.A., 1999. Sleep and wakefulness modulation of the neuronal firing in the auditory cortex of the guinea-pig. Brain Res 816, 463–470.

Pérez-Perera, L., Bentancor, C., Pedemonte, M., Velluti, R.A., 2001. Auditory cortex unitary activity correlated to sleep-wakefulness and theta rhythm in response to natural sounds. Actas Fisiología 7, 187.

Portas, C.M., Krakow, K., Allen, P., Josephs, O., Armony, J.L., Frith, C.D., 2000. Auditory processing across the sep-wake cycle: simultaneous EEG and fMRI monitoring in human. Neuron 28, 991–999.

Terzano, M.G., Parrino, L., Fioriti, G., Orofiamma, B., Depoortere, H., 1990. Modifications of sleep structure by increasing levels of acoustic perturbation in normal subjects. Electroenceph. Clin. Neurophysiol. 76, 29–38.

Vallet, M., 1982. La perturbation du sommeil par le bruit. Soz. Praventivmed. 27, 124–131.

Vallet, M., Mouret, J., 1984. Sleep disturbance due to transportation noise: ear plugs vs oral drugs. Experientia 40, 429–437.

Van Twyvert, H.B., Levitt, R.A., Dunn, R.S., 1966. The effect of high intensity white noise on the sleep pattern of the rat. Psychon. Sci. 6, 355–356.

Velluti, R.A., 1997. Interactions between sleep and sensory physiology. A review. J. Sleep Res. 6, 61–77.

Velluti, R.A., 2005. Remarks on sensory neurophysiological mechanisms participating in active sleep processes. In: Parmeggiani, P.L., Velluti, R.A. (Eds.), The Physiologic Nature of Sleep. Imperial College Press, London, pp. 247–265.

Velluti, R.A., Pedemonte, M., 2012. Sensory neurophysiologic functions participating in active sleep processes. Sleep Sci. 5 (4), 103–106.

Velluti, R.A., Roig, J.A., Escarcena, L.A., Villar, J.I., García Austt, E., 1965. Changes of brain pO2 during arousal and aletness in unrestrained cats. Acta Neurol. Latinoamer. 11, 368–382.

Velluti, R.A., Pedemonte, M., Suárez, H., Inderkum, A., Rodríguez-Servetti, Z., Rodríguez-Alvez, A., 2003. Human sleep architecture shifts due to auditory sensory input. Sleep 26 (Suppl.), A19.

Velluti, R.A., Pedemonte, M., Suárez, H., Bentancor, C., Rodriguez-Servetti, Z., 2010. Auditory input modulates sleep: an intra-cochlear implanted human model. J. Sleep Res. 19 (4), 585–590.

CHAPTER 7

Tinnitus Treatment During Sleep

Marisa Pedemonte
CLAEH University, Punta del Este, Uruguay

INTRODUCTION

Tinnitus

Subjective idiopathic tinnitus is a widespread disabling condition, with a prevalence estimated at 10%–15% (Henry et al., 2005), seriously affecting the quality of life in 1%–2% of the population (Pilgramm and Rychalik, 1999). Mechanisms that underlie tinnitus perception are still not well understood; nevertheless, nowadays it is accepted that interactions between altered cochlear inputs and distorted central auditory processing provoke tinnitus. The physiological abnormalities that cause subjective tinnitus perception arise in the central nervous system (CNS). It is now evident that most forms of subjective tinnitus are caused by changes in the function of the central auditory nervous system, while these changes are not associated with any detectable anatomical lesion. Subjective tinnitus may be the result of the expression of neural plasticity and the anomalies may develop because of decreased input from the ear, deprivation of sound stimulation, overstimulation or yet unknown factors (Jastreboff, 1990). Studies with brain imaging support the central correlates of tinnitus perception (Andersson et al., 2000; Melcher et al., 2000). In a number of patients subjective tinnitus may be initiated by a discontinuity in the spontaneous or low-level stimulus, inducing neural activity across auditory nerve fibres with different characteristic frequencies (CFs). This discontinuity may be caused by functional loss of outer hair cells in those regions where inner hair cells are preserved. The reduced spontaneous activity for nerve fibres with CFs in the hearing loss range may result in a reduction of inhibition, mediated by the auditory efferent system, at more central levels (Eggermont and Roberts, 2004). This reduced inhibition of neurons with CFs induces hypersensitivity and hyperactivity in these neurons, generating a 'phantom sensation'. Because the brain is not able to discern if this abnormal incoming flux of information is related to real

The Auditory System in Sleep
DOI: https://doi.org/10.1016/B978-0-12-810476-7.00007-5

environmental sound, a 'phantom sensation' (tinnitus) may be created (Jastreboff, 1990). Hyperactivity may also be the result of reorganization of the cortical tonotopic map after cochlear damage, which induces a release from efferent inhibition at the CF that lose cortical representation (Eggermont and Komiya, 2000; Noreña and Eggermont, 2003; Rauschecker, 1999; Robertson and Irvine, 1989). Increase in the spontaneous firing rate has been found at different levels of the auditory pathway: dorsal cochlear nucleus (Brozoski et al., 2002), inferior colliculus (Mulders and Robertson, 2009) and auditory cortex (Noreña and Eggermont, 2005). Since most forms of severe tinnitus are caused by functional changes, our hypothesis was that it should be possible to reverse it with proper sound treatment, taking advantages of the neural plasticity properties of the CNS.

Auditory System and Sleep–Waking Cycle Interactions

In the preceding chapters the interaction between auditory processing and the wake–sleep cycle has been detailed. Here, we present the main antecedents that support the development of tinnitus treatment.

Auditory Processing During Sleep

While all sensory processing persists during sleep, the auditory input is particularly relevant for continuously monitoring the environment. Phylogenetically it has developed like a sentinel to protect against the predators, to attend the danger it claims, or to decide that sleeper can continue sleeping, if it considers that the sound stimulus received is inconsequential, among other roles (Velluti, 1997, 2008).

There is a consensus that sleep is involved in the processes of learning and consolidation of memory, but it is not yet clear what role each stage of sleep plays in different types of memory. It has been argued that slow wave sleep is important for declarative memory and that working memory is processed mainly during rapid eye movements (REMs) sleep; however, the integrity and interaction of the different sleep stages for learning and memory process has also been suggested (Cipolli, 2005; De Gennaro et al., 2000; Mölle and Born, 2011; Mölle et al., 2011). Slow electroencephalography (EEG) oscillations (less than 1 Hz) have been involved in the consolidation of long-term memory (Diekelmann and Born, 2010) and in the homoeostatic regulation of synaptic connections (Tononi and Cirelli, 2006). Rhythmic acoustic stimulation induces K-complexes, which are considered a 'forerunner' of slow oscillations in slow wave sleep stage (De Gennaro et al., 2000; Riedner et al., 2011). Slow oscillation during slow wave sleep promotes consolidation of memory and the post-sleep facilitation of encoding new memories (Marshall et al., 2006). Slow waves may be modulated by low frequency auditory stimulation (Ngo et al., 2013). Studies with functional magnetic resonance showed that auditory cortical activity is maintained during sleep but varies with stimulus significance (Maquet et al., 2005; Portas et al., 2000).

Auditory Input Modulates Sleep

The auditory total deprivation, by surgical bilateral cochleae lesions in the guinea pig, modified the wakefulness/sleep cycle, diminishing the total wakefulness duration while increasing slow wave sleep, paradoxical sleep and the number of episodes of each behavioural state in guinea pigs (Pedemonte et al., 1996). Furthermore, noisy ambient and sound stimulation may also alter the sleep structure (Vallet, 1982; Drucker-Colín et al., 1990; Ball et al., 1989). Both hamsters' biological rhythms, motor activity (a circadian rhythm) and wakefulness–sleep cycle (an ultradian one) changed when the animal was auditory deprived (deaf) by cochlear lesion. The most important change found in the wakefulness–sleep rhythm was the incremented number of episodes and, in some cases, the duration of sleep and wakefulness, demonstrating the auditory deprivation altered the normal waking–sleep architecture (Cutrera et al., 2000). This result is in agreement with those previously found in deaf guinea pigs (Pedemonte et al., 1996).

On the other hand, human imaging techniques reported low levels of auditory cortical activity among deaf subjects: The longer is the duration of deafness, the lower the level of activity recorded. Moreover, the metabolic activity in the primary auditory cortex was observed to increase bilaterally to nearly normal levels after a successful cochlear implantation, although the greatest activity was obtained using functional magnetic resonance imaging on the side contralateral to the implant (Lazeyras et al., 2002). Maintenance of intracochlear implant at 'on' during sleep improves sleep quality and slow wave stages (Velluti et al., 2010).

There is currently no doubt that the interactions between auditory processing and sleep are complex and reciprocal (Velluti and Pedemonte, 2012). Given all this background of the interaction of sound stimulation with different stages of sleep, we have explored the changes in brain activity induced by sound stimulation during sleep in patients with tinnitus (Pedemonte et al., 2014).

Tinnitus and Sleep Disorders

Studies demonstrated a strong relationship between sleep disorders and tinnitus. A large percentage of patients with tinnitus experience sleep disorders; poor sleep and frequent waking are more common among subjects with tinnitus (60%). Sleep disturbances are a factor that strongly predicts decreased tolerance to tinnitus (Eysel-Gosepath and Selivanova, 2005). Furthermore, patients whose sleep was most disturbed rated significantly greater tinnitus annoyance in the evening, underscoring the influence of tinnitus on sleep disorders (Hallam, 1996). On the other hand, daytime sleepiness is more common in subjects with tinnitus and it is even more frequent in patients with both tinnitus and poor sleep (Asplund, 2003). Insomnia is associated with greater perceived loudness and severity of tinnitus (Folmer and Griest, 2000). These facts underscore the importance of identification and successful treatment of sleep disorders in patients with tinnitus.

Tinnitus Treatments

The treatment of idiopathic tinnitus has always been a great medical challenge, since partial or sustained healing or remission of this symptom has not yet been found. However, the greater knowledge of the pathophysiology that triggers tinnitus, placing it as an error in auditory processing at the level of the CNS, has evolved the therapeutic protocols to more rational schemes. The history of treatments has included, for example, the destruction of the cochlea or the section of the auditory nerve that was realized decades ago, when the central genesis of the process was unknown. Fortunately, these treatments have long been deprecated, since they left the patient totally deaf and increased the imbalance of the input of auditory information, which also made tinnitus worse.

When the patient has a concomitant hearing loss with indication of hearing aids, often improving the auditory input balance is sufficient to improve tinnitus (König et al., 2006). However, in most cases, especially in young subjects, although there is an alteration in the audiogram with asymmetry (Fig. 7.1), this does not justify the indication of hearing aids.

Pharmacological Treatments and Others

The knowledge that tinnitus is a product of the interaction of auditory dysfunction, cognitive and attention changes, and emotional aspects such as anxiety and depression has led to the development of several paradigms of

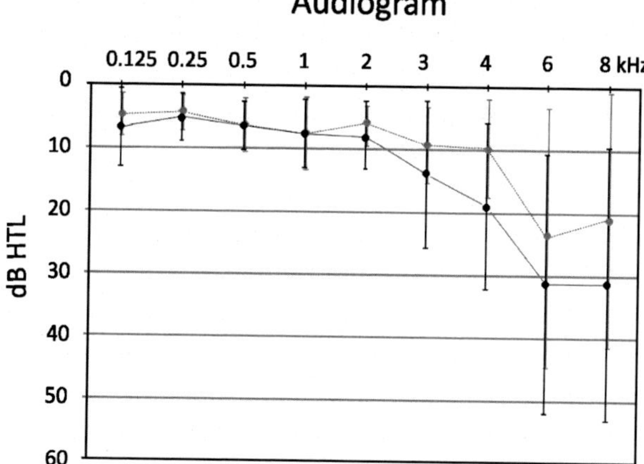

Figure 7.1 Mean audiogram of 11 patients from the second clinical trial. For each frequency a mean of liminal thresholds and their standard deviation is plotted. *Modified from Drexler, D., López-Paullier, M., Rodio, S., González, M., Geisinger, D., Pedemonte, M., 2016. Impact of reduction of tinnitus intensity on patients' quality of life. Int. J. Audiol. 55(1), 11–19.*

treatment, such as teaching to differentiate between tinnitus and other sounds, improving the ability to discriminate, psychological treatments, relaxation techniques, cognitive-behavioural therapies, stress reduction, exercises, physiotherapy, acupuncture and electroacupuncture, among others.

Pharmacological therapies are a tool commonly used by the physician, where various drugs are used (lidocaine i/v, benzodiazepines, baclofen, carbamazepine, ginkgo biloba, nimodipine, dihydroergotoxina and idebenone). Drugs have not demonstrated the ability to provide replicable long-term reduction of tinnitus (Langguth et al., 2009).

Other treatments that are still in research stages include electrical stimulation at the cochlear level or through deep implants in different nuclei from the auditory pathway to the primary auditory cortex. The possible benefit of transcranial magnetic stimulation is also being studied (de Ridder et al., 2007; Kleinjung et al., 2007). Many of these are still at a research level, such as transcranial magnetic or cortical electric stimulation of the auditory cortex.

Tinnitus Treatment with Sound Stimulation

Several treatments for tinnitus are based on sound stimulation at present, such as tinnitus masking, music and white noise. Tinnitus retraining therapy (TRT) is a behavioural treatment that has roots in the hypothesis that tinnitus is caused by expression of the neural plasticity (Jastreboff, 1995; von Wedel et al., 1997; McKinney et al., 1999). The aim of TRT is to psychologically disconnect the patient from dependence on the tinnitus while subjecting the patient to moderate levels of sounds to reverse the effect of sound deprivation on the function of the CNS (Kroener-Herwig et al., 2000). Other tinnitus therapies based on sound being applied are a special programme of music therapy that strives to integrate the tinnitus sound into a musically controllable acoustic process (Kusatz et al., 2005), sequential sound therapy (López González and López Fernández, 2004) and customized sounds (Pineda et al., 2008; Hanley and Davis, 2008). Many of these therapies are still under investigation, and only some of them have been objectively evaluated in clinical trials (Jastreboff, 2007). Up to date, no other therapies have shown to improve the tinnitus with certainty.

Sound Stimulation Mimics the Tinnitus

The evolution of the research has shown that the best results are those involving sound stimulation that one way or another takes into account some of the characteristics of tinnitus. The use of customized acoustic stimuli is supported by studies (Noreña and Eggermont, 2005; Schaette et al., 2010). Varied sounds or melodies with modifications in their component frequencies, with added white noise, pure tones with phase shift, amplitude and frequency modulation are some of the stimulating sound modulation strategies (Pantev et al., 2012; Wazen et al., 2011; Vermeire et al., 2007; Wilde et al., 2008; Reavis et al., 2012; Heijneman et al., 2012).

Sound Stimulation Applied During Sleep with Sound That Mimics Tinnitus

Based on the knowledge that auditory processing continues during sleep (Velluti, 2008) and that a relationship between learning, memory and sleep stages has been established, our group, in Uruguay, embarked on a new strategy for the treatment of idiopathic subjective tinnitus. We demonstrated effectiveness in decreasing the intensity of idiopathic subjective tinnitus and concomitantly improving the quality of life of patients suffering from it, through two clinical trials (Pedemonte et al., 2010; Drexler et al., 2016).

Patient Evaluation

Inclusion and Exclusion Criteria

Sixty-six patients suffering from tinnitus were evaluated in two clinical trials (Pedemonte et al., 2010; Drexler et al., 2016), of which 22 participants were selected according to the criteria of inclusion and exclusion. The inclusion criteria were patients from 18 to 70 years old with unilateral or bilateral subjective idiopathic tinnitus, experiencing tinnitus for more than 6 months and a tinnitus handicap inventory (THI) score above 17. The exclusion criteria included patients that demonstrated objective or subjective secondary tinnitus, hearing loss of 50 dB or less on the hearing threshold level (HTL) in more than two frequencies of the audiogram, patients that had undergone other treatments for tinnitus in the past year, current use of hearing aids, use of psychoactive drugs, depression (Hamilton scale test above 13) and sleep disorders other than those provoked by the tinnitus itself. Patients with sleep disturbances such as apnoea, restless legs syndrome, narcolepsy, parasomnia and insomnia with aetiology other than tinnitus were excluded from the sample.

To evaluate these criteria all patients were interviewed and examined by an otolaryngologist, an audiologist and a psychologist. The laboratory profile consisted of imaging studies (MRI or CT scan) and blood tests (blood lipids, thyroid hormones, glucose tests, urea and electrolytes and creatinine).

Audiologic Evaluation

Audiometric profiles were performed using impedanciometry, audiometry, loudness discomfort levels, speech audiometry and high frequency audiometry. Three types of otoacoustic emissions were measured: distortion product otoacoustic emissions, transient evoked otoacoustic emissions and spontaneous otoacoustic emissions. Fig. 7.1 shows the average of audiograms of patients in the second clinical trial.

Sleep Evaluation

Patients' sleep conditions were clinically evaluated, through a clinical interview with a physician who specialized in sleep. At the beginning of the first clinical trial, a conventional polysomnography was performed in four patients, for two nights each (the first night without sound stimulation and the second during stimulation with its specific sound) to detect possible hypnogram changes due to stimulation of the sound. Conventional polysomnography consisted of six electroencephalographic

Table 7.1 Tinnitus functional index

Subscales	Control	Middle test (6 weeks)	Last test (12th week)
Intrusive (unpleasantness, intrusiveness and persistence)	51.4	33.2*	34.9*
Sense of control (reduced)	67.0	41.3*	30.8*
Cognitive (cognitive interference)	32.1	19.1	8.2**
Sleep (sleep disturbance)	68.4	28.2*	13.3*
Auditory (auditory difficulties attributed to tinnitus)	29.8	11.3*	10.5*
Relaxation (interference with relaxation)	64.2	29.7*	11.5**
Quality of life (reduced)	35.9	13.9*	5.1*
Emotional (emotional distress)	51.2	18.5*	12.5*

Note: Tinnitus functional index subscales were analysed separately; each number corresponds to the average of 11 patients. Asterisks identify statistically significant changes between Control versus Middle and Control versus Last test (*$P<.05$; **$P<.001$).
Source: Wilcoxon test, from Drexler, D., López-Paullier, M., Rodio, S., González, M., Geisinger, D., Pedemonte, M., 2016. Impact of reduction of tinnitus intensity on patients' quality of life. Int. J. Audiol. 55(1), 11–19.

recordings, eye movements, electromyogram, respiration, oximetry and electrocardiogram. No changes were found either in the percentages or in the latencies of the sleep stages nor in the temporal configuration of their appearance at night. Neither were changes found in sleep efficiency or in the number and duration of awakenings and arousals (Pedemonte et al., 2007). During treatment development sleep quality was evaluated clinically in each interview. Subjective perception about the time to sleep onset, maintenance of sleep and early waking were explored. The repercussion on wakefulness, such as loss of memory, irritability and somnolence, were also evaluated. The impact of tinnitus on sleep quality was evaluated through the tinnitus functional index's (TFI) sleep subscale test. In the subsequent clinical trial, this result was reaffirmed, showing that the sleep disturbance subscale of the TFI improved with statistical significance (Table 7.1; Drexler et al., 2016).

Psychological Evaluation

Evaluation by a psychologist was performed at the beginning, in the middle (week 6) and at the end of the treatment (week 12). In each interview, patients were specifically evaluated for anxiety and depression. The impact of tinnitus on quality of life was assessed through three

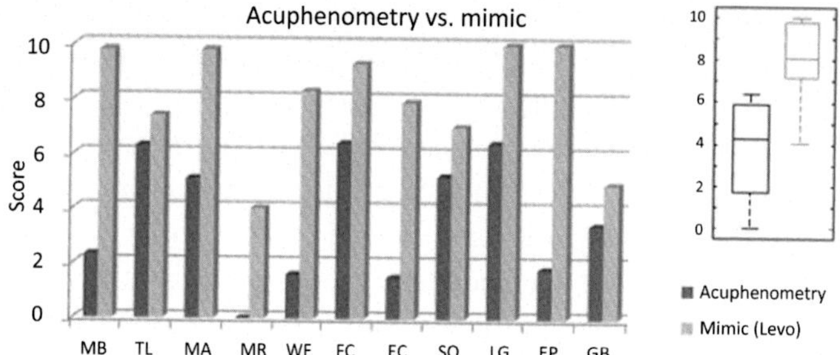

Figure 7.2 Comparison between the acuphenometry and the sound created to mimic the tinnitus for nocturnal stimulation (11 patients, from the second clinical trial). Each patient set the similarity of these sounds with his or her tinnitus, through a visual analog scale ($P < .05$; Wilcoxon test).

questionnaires: the THI, tinnitus reaction questionnaire (TRQ) and TFI. A visual analog scale (VAS) regarding tinnitus annoyance was also performed at each interview.

Tinnitus Characterization and Sound Stimulation

Tinnitus Sound Match

Evaluation of the sound match accuracy was performed a week after the customized sound was first created. Patients' feedback was requested on the capacity of the sound to mask their tinnitus. The sound was changed if it did not achieve that goal. Once masking was verified, VAS on the similarity to the tinnitus was conducted to obtain a customized sound that not only masked the tinnitus but also was as similar as possible to the patient's perception. This latter point was important because the more similar the sound match was to the patient's perception, the less intensity was required to mask it and therefore the more accurate were the intensity measurement and the intensity follow-up. In the VAS the patient was asked to rank the matching between 0 (totally different from tinnitus) and 10 (exactly the same as the perceived tinnitus). A sound match that scored over 6 in the VAS scale and covered the patient's perception was accepted to start the treatment. The average VAS score for tinnitus match in this study was 8.3 while average VAS score for acuphenometry was 4.1. The difference observed in these scores was statistically significant (Fig. 7.2).

Generation of Stimulating Sound

Taking into account that most tinnitus reported by patients was classified as complex sounds (Henry, 2004), we designed software with the specific aim of matching the patients' perception by a highly customized combination, with the capability of presenting the patient five types of sounds: pure tones, bandpass noise, a 'cricket' sound, white noise and pink noise. This software was loaded onto the physician's device (iPad). Separate software, which was capable of reproducing the customized sound, was loaded onto the patients' devices (iPod Touch). These devices allowed patients to set the intensity of sound stimulation every night and store sound-intensity data daily. These two programs (one running on the physician's iPad and the other on the patients' iPods) were designed to be able to communicate via Wi-Fi, sharing data of intensity evolution and daily time of stimulation. Using this capability, information stored on the patient's iPod could be downloaded to the physician's iPad during each appointment and displayed graphically. Customized in-ear ear buds were created for each patient. Solid soft-gel medical grade silicon moulds were made based on impressions taken from the ear canals. The moulds were manufactured to achieve maximum occlusion, in order to avoid air leaks between them and the walls of the ear canal. They were also designed to not protrude from the ear's pinna. The latter two features were important to prevent the customized sound from being heard by surrounding people and to enable the patient to comfortably lay his or her ears on the pillow while sleeping. The occlusive moulds did not move once they were located in the ear canal, staying in place during sleep and thus ensuring a fixed distance between the source of sound and the eardrum. This point was fundamental for achieving stable and calibrated sound pressure in the ear canal. The response curve of all the customized ear buds used in this study was characterized and a mean response curve was derived. For more details about sound generation and methods, see Drexler et al. (2016) and Patents US 9.282.917 B2 and US 9.301.714 B2.

CLINICAL RESULTS

Tinnitus Intensity Reduction

Two clinical trials were conducted in patients with idiopathic subjective tinnitus. In the first trial the objective was to evaluate the existence of a measurable and quantifiable reduction of tinnitus intensity in patients who were followed for 6 months of treatment (Pedemonte et al., 2010). The results showed that the main fall in intensity occurred in the first 2 weeks, with a slow and steady slope of descent in the remaining 22 weeks. In the second clinical trial patients were followed for 12 weeks (3 months), where the results of the first trial were repeated (Fig. 7.3).

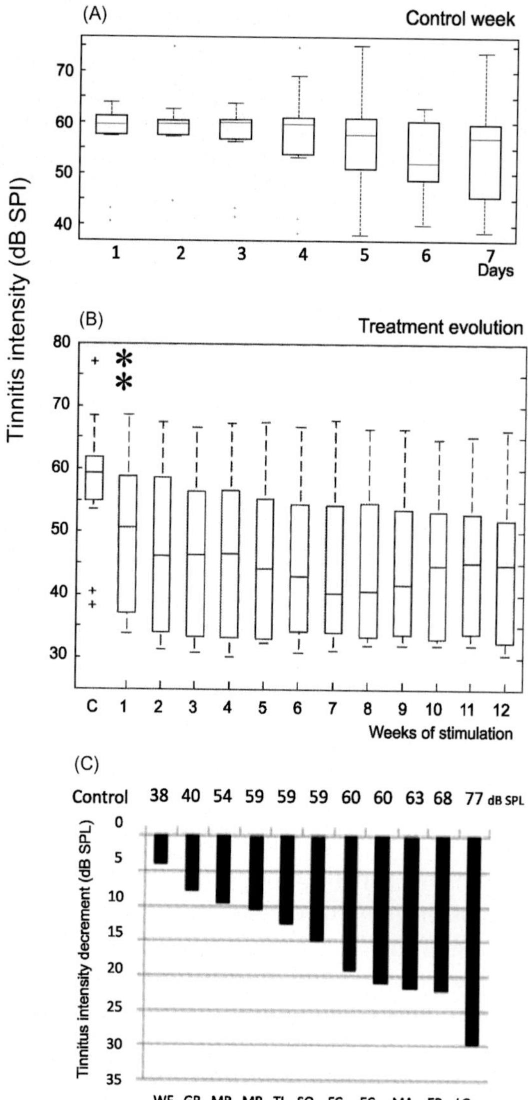

Figure 7.3 Evolution of tinnitus intensity through 12 weeks. (A) Box plot showing averages of tinnitus intensity measured during 7 days without stimulation. Each box represents the daily average for 11 patients. No statistically significant changes were observed between the tinnitus intensity averages ($P = .09$, ANOVA). (B) Box plot showing the average weekly intensities for the same patients during 3 months of stimulation during sleep. *C* is the 'control' considering the average across 7 days of tinnitus intensity measurements prior to the beginning of stimulation. On average, tinnitus intensity decreased 14.1 dB between Control and week 12. These results are statistically significant (**, $P < .001$) comparing values between Control versus the 1st week and Control versus the 12th week (Wilcoxon test). (C) Decrease of tinnitus intensity. Bars show intensity reduction for each patient. Zero corresponds to the measurement of tinnitus intensity before treatment (average of 7 days); the range of decrease was between 3.9 and 29.8 dB, taking into account the values of the 12th week of treatment. At the top (Control), initial levels of tinnitus intensity in each patient are shown. *Modified from Drexler, D., López-Paullier, M., Rodio, S., González, M., Geisinger, D., Pedemonte, M., 2016. Impact of reduction of tinnitus intensity on patients' quality of life. Int. J. Audiol. 55(1), 11–19.*

After 3 months of treatment, a mean decrease in tinnitus intensity of 14.1 dB SPL was observed, implying a 62% reduction in perceived sound. It is important to note that, although the degree of improvement was different in each patient, none of their tinnitus was worsened by the treatment. Psychological follow-up by personal interviews showed that there was no strict relationship between the decrease in tinnitus intensity and improvement in the patient's quality of life, so some with small falls felt much better than others with significant improvements in the decrease of tinnitus (Drexler et al., 2016). The average tinnitus intensity over all 11 patients showed no statistically significant difference during the 7 days before the beginning of stimulation ('control week', Fig. 7.3A). The intensity values obtained from the 'control week' were averaged to obtain the control level prior to starting treatment. Each patient showed a statistically significant decrease in tinnitus intensity, comparing seven values collected in the control week (Fig. 7.3A, mean 58 dB SPL), with seven values in week 1 and the control week versus week 12 (Fig. 7.3B, mean 43.9 dB SPL), that is 14.1 dB SPL. Statistical significance was evaluated using the Wilcoxon signed-rank test ($P<.001$, nonparametric, paired samples). The 11 patients studied showed a decrease of the tinnitus intensity that varied from a maximum of 29.8 dB to a minimum of 3.9 dB (Fig. 7.3C).

During the study, no patient experienced an increase in the intensity of his or her tinnitus; in none of them had the tinnitus completely disappeared in a maintained form, although some experienced tinnitus disappearance for hours or days. The degree of improvement was not associated with the sound characteristics of tinnitus or the intensity during the control week. Also, no significant correlation was observed between the effect of treatment and the time evolution of tinnitus.

Psychological Evaluation of Patients

The results of the three psychometric tests performed showed statistically significant improvements. The biggest improvement was between the first and second rounds of tests (the first 45 days of treatment) with an average decrease of 27.3 for THI ($P<.001$), 39.0 for TFI ($P<.001$) and 24.7 for TRQ ($P<.01$, Fig. 7.4). During the third round of tests (at the end of the third month of treatment), the results still showed improvement in comparison to the second round, however, these changes were not statistically significant. The analysis of TFI showed statistically significant

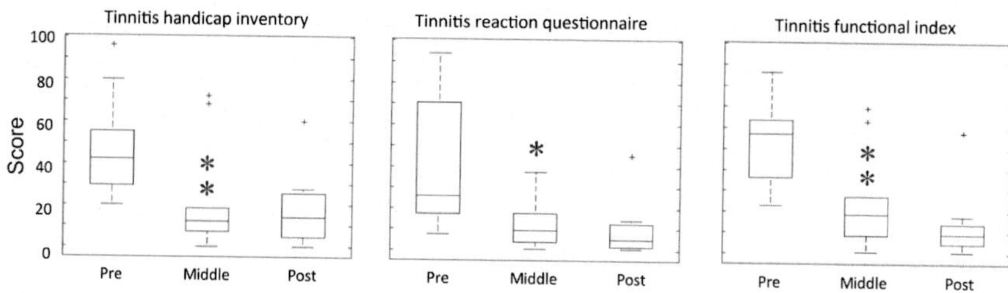

Figure 7.4 Psychometric evolution of patients. Psychometric assessment was conducted using three tests: tinnitus handicap inventory, tinnitus reaction questionnaire and TFI. These were performed before starting treatment (Pre), and in the 6th week (Middle) and 12th week (Post) after starting treatment. Statistically significant differences were found between (Pre) and (Middle) values, in all three tests (THI and TFI, **, $P < .001$ and TRQ, *, $P < .01$, Wilcoxon test). *From Drexler, D., López-Paullier, M., Rodio, S., González, M., Geisinger, D., Pedemonte, M., 2016. Impact of reduction of tinnitus intensity on patients' quality of life. Int. J. Audiol. 55(1), 11–19.*

decreases between the first and second rounds for the intrusive, sense of control, sleep, relaxation, emotional and quality of life subscales. Also, significant changes where observed between the first and third rounds of tests for all the subscales (Table 7.1), with especially relevant improvements in sleep and relaxation, which had reductions of 55.1 and 52.7, respectively. Sleep was also evaluated clinically in all patients in each interview, showing an improvement in their quality and quantity of sleep. Patients reported falling asleep faster, waking up fewer times during the night or experiencing longer sleep times.

A VAS was used to determine and quantify the level of annoyance induced by the tinnitus. The VAS was performed before starting the treatment (Pre), in the 6th week (Middle) and 12th week (Post). All the patients showed a reduction in their tinnitus annoyance. The average result of the scores was 6.53 for the preround of tests, 3.88 and 2.55 for the middle and postrounds, respectively. An overall improvement of 61% of the pretreatment scores was observed. The difference between pre- and middle treatment average scores was statistically significant (Wilcoxon: $P < .001$).

The Impact of Sound Stimulation During Sleep on Electroencephalographic Waves

The study of sleep during sound stimulation had two objectives: (1) as mentioned at the beginning, to demonstrate that the application of stimulating sound did not affect any of the sleep parameters (quality, quantity, temporal configuration, opportunity, etc.) or repercussions in the day; and (2) try to understand the changes that underlie sound stimulation in brain activity to begin to interpret the mechanisms that are activated to improve tinnitus (Pedemonte et al., 2014).

Ten patients from the second clinical trial were studied with the polysomnography that was carried out with the usual clinical protocol recording 10 electroencephalographic channels (EEG, frontals, F3, F4; centrals C3, C4; parietals P3, P4; temporal T3, T4, T5, T6, following the internationally accepted standard denomination), electrocardiogram, electromyogram, eye movements and oxygen saturation. All EEG recordings were monopolar, recorded from scalp electrodes and separate ear electrodes A1 and A2, with electrodes referenced to linked ear lobes. The sampling frequency was set at 256 Hz. The EEG acquisition system is equipped with hardware high-pass filters with cutoff frequency at 0.5 Hz and hardware low-pass filters with cutoff frequency at 100 Hz. Also there is a selectable notch filter to suppress 50/60 Hz power line noise. No digital postprocessing filters were applied. One researcher accompanied the patient all night, diagnosing the sleep stages online.

After beginning the night with the usual sound stimulation for tinnitus treatment, sound is stopped after a minimum of one pass through each of the sleep stages: somnolence (stage N1), stage N2 with sleep spindles, stage N3 with slow wave sleep and sleep with REM. The rest of the night the patients continue to sleep in silence (Fig. 7.5). All patients started with sound stimulation because they are habituated to it and the sound improves sleep onset. This enhances the disturbances caused by tinnitus, such as anxiety, increased sleep latency, awakenings and shallow sleep the first few hours. Twenty temporal windows (2 second duration each one) were selected in each sleep stage (N2, N3 and REM); 10 of them during silence and the other 10 during sound stimulation. Always data were compared in the same patient. The power spectra and the coherence in electroencephalographic waves recorded by electrodes F3, F4, T3 and T4 were analysed. We compared the power spectra during noiseless (as a control) versus sound stimulation, exploring different electroencephalographic frequency bands (delta, 0.5–3.5 cps; theta, 4–7.5 cps; alpha, 8–12 cps) in the same sleep stage. A comparison between the left and right hemispheres (T3 vs T4 and F3 vs F4) was also carried out. We studied the wave's coherence percentages, analysing a pair of intrahemispheric electrodes (F3–T3 and F4–T4) and interhemispheric electrodes (F3–F4 and T3–T4). The overall coherence (considering all

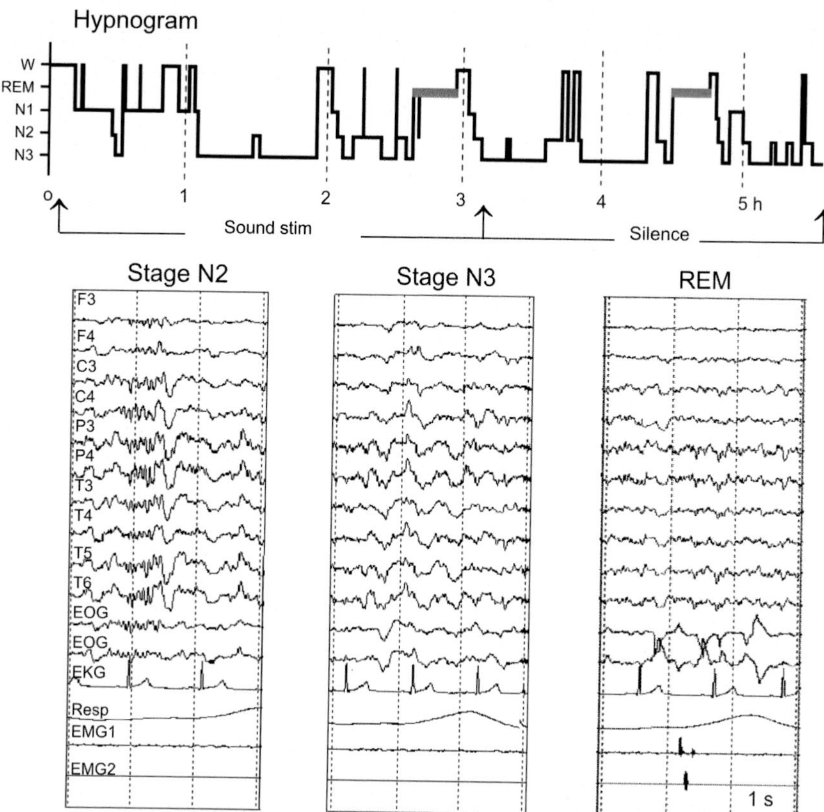

Figure 7.5 Polysomnographic recording of patient L.G. She was recorded during nocturnal physiological sleep for almost 6 h. From top to bottom, hypnogram, showing sequences of the different sleep stages through the night: wakefulness (W), rapid eyes movement sleep (REM) and stages N1, N2 and N3. Sound stimulation was applied in the first half of the night (arrows). The three boxes show 3 s of different sleep stages: 10 electroencephalographic recordings, 2 electrooculograms (EOG), electrocardiogram (EKG), respiratory movements (Resp) and legs movements (EMG1 and EMG2). *From Pedemonte, M., Testa, M., Díaz, M., Suárez-Bagnasco, D., 2014. The impact of sound on electroencephalographic waves during sleep in patients suffering from tinnitus. Sleep Sci. 7, 143–151. Open access, online: doi:10.1016/j.slsci.2014.09.011.*

frequencies, from 0.5 to 12 cps, together) and the coherence of each range of frequency were considered, comparing temporal windows during silence with temporal windows during sound stimulation. Signal processing was done using sets of 10 time blocks for each measurement channel. The duration of each time block was 2 seconds (see details of the coherence processing algorithm in Pedemonte et al., 2014).

Power Spectra Analysis

Sound stimulation induced changes in the power spectra of EEG waves during all sleep stages. Regarding sleep stages, 35.5% of statistically significant changes took place during stage N2, 35.5% in stage N3 and the remaining 29% during REM sleep. The 27% of the significant changes were decrements while the 73% showed increments in the power spectrum during the sound stimulation. Across all patients and stages, the theta was the band that showed most significant changes (48%), followed by the delta band (36%). The alpha band displayed the smallest change (16%). Considering recording location, temporal electrodes showed larger change than frontal ones (61% vs 39%, respectively). Fig. 7.6 shows an example of the analysis protocol conducted on each dataset. Thirty-six comparisons were performed across 10 temporal windows selected in each sleep stage. The power spectra of delta, theta and alpha bands were analysed by comparing the periods of silence and sound stimulation at frontal and temporal electrodes (F3, F4, T3 and T4). In this example (Fig. 7.6), 27 out of 36 power spectra showed increment (75%), 5 of them being statistically significant; only 1 interhemispheric comparison was also statistically significant (in REM sleep) in this patient binaurally stimulated (Pedemonte et al., 2014).

Coherence Analysis

Statistically significant changes were found in both intrahemispheric and interhemispheric coherence in every sleep stage and in all the frequency bands analysed comparing coherence control (during silence) with periods with sound stimulation. The inter- and intrahemispheric differences that appeared with sound stimulation were not present in the control situation. All patients also showed some sort of differences in coherence intra- and interhemispheric in control situation, regardless whether the tinnitus was uni- or bilateral. These differences varied depending on the stage of sleep considered and the range of bands analysed.

During sound stimulation most of the changes in coherence occurred in the N2 sleep stage, both when considering overall coherence as well as the different bands separately; and stage N3, being in REM sleep, showed the least amount of changes. In stage N2 the spindles showed most of the changes, followed by bands theta, delta and alpha in descending order. In stages N3 and REM most of the changes appeared in the delta band, followed by theta and alpha bands.

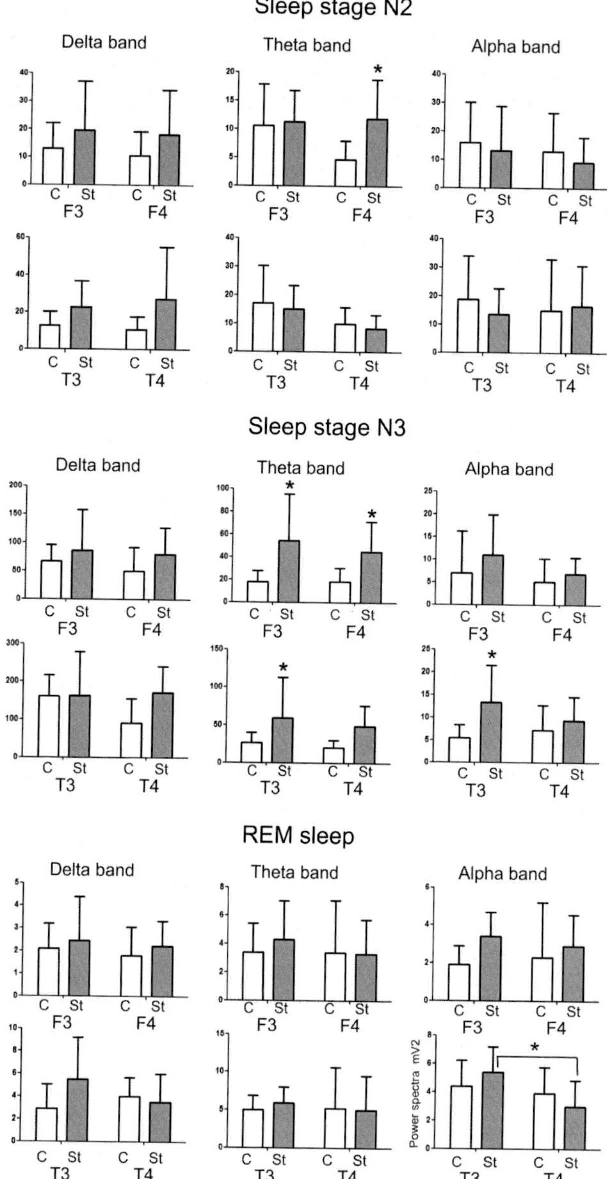

Figure 7.6 Power spectra of patient T.L. binaurally stimulated. Power spectra of delta, theta and alpha electroencephalographic bands were studied in frontal (F) and temporal (T) electrodes (3 is the left side, 4 is the right side), in each sleep stage (N2, N3 and REM). Ten temporal windows during sound stimulation (St) were compared with another 10 in silence as a control (C) in each situation. Bars show means ± standard deviation. Data were statistically analysed by one-way ANOVA followed by Tukey posttest to compare all pairs of columns, $*P < .05$. *From Pedemonte, M., Testa, M., Díaz, M., Suárez-Bagnasco, D., 2014. The impact of sound on electroencephalographic waves during sleep in patients suffering from tinnitus. Sleep Sci. 7, 143–151. Open access, online: doi:10.1016/j.slsci.2014.09.011.*

Changes in the N2 stage appeared both in the sense of the increase and the decrease of coherence for the delta, theta and alpha bands; however, the spindles changed, reducing coherence most of the time. During sound stimulation in stage N3, coherence decreased for all studied bands (delta, theta and alpha). When the stimulation occurred during REM sleep, the coherence increased two-thirds of the time, decreasing in the remaining third, for all the bands studied. Fig. 7.7 shows changes of coherence in a patient with unilateral tinnitus, stimulated at the right side. In this example changes appear only in stage N2 with the sound stimulation. Intrahemispheric overall coherence decreased in both hemispheres, and spindles only decreased at the right. There are differences in

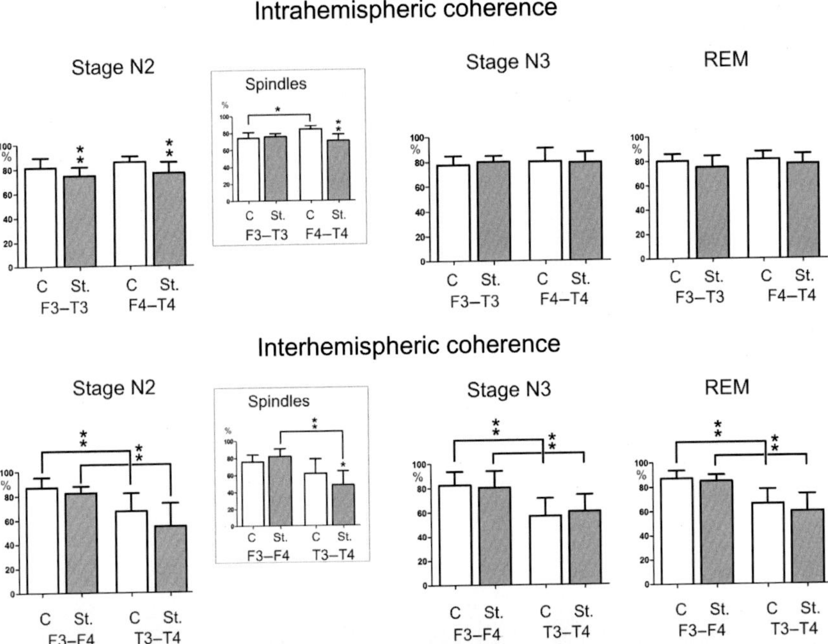

Figure 7.7 Overall coherence percentages in patient E.C. with monaural stimulation (right side). Overall coherence percentages were analysed in intrahemispheric electrodes (F3–T3 and F4–T4) and interhemispheric electrodes (F3–F4 and T3–T4) during stage N2, N3 and REM sleep. Inserts, coherence of frequencies that make up the sleep spindles in stage N2 (13–16 Hz) also were studied. Bars show mean ± standard deviation. Data were statistically analysed by nonparametric ANOVA (Kruskal–Wallis test) and post hoc Dunn's Multiple Comparisons test, $*P<.05$, $**P<.01$. *From Pedemonte, M., Testa, M., Díaz, M., Suárez-Bagnasco, D., 2014. The impact of sound on electroencephalographic waves during sleep in patients suffering from tinnitus. Sleep Sci. 7, 143–151. Open access, online: doi:10.1016/j.slsci.2014.09.011.*

the spindle coherence between the right and left hemispheres during the silence; these differences are not maintained during sound stimulation (top inset). Interhemispheric coherence shows differences in the spindles between temporal and frontal areas only during sound stimulation. The differences between controls at all sleep stages were kept during stimulation (Pedemonte et al., 2014).

Intrahemispheric Coherence

Intrahemispheric changes, compared between periods of sound stimulation and silence, appeared mainly in the left hemisphere, regardless of the stimulated side (binaural, right or left; F3–T3, 65% vs F4–T4, 35%).

Interhemispheric Coherence

Temporal electrodes (T3–T4) showed more changes of coherence than frontal ones (F3–F4) in response to sound stimulation (72% vs 28%). Four patients without differences between frontal and temporal coherence during control showed differences when sound stimulated. Most of the interhemispheric coherence changes appear in stage N2 (53%) followed by stage N3 (37%) and REM (10%). Taking account of the different bands, the delta band was changed the most in all sleep stages (67%), with predominance in temporal areas (47%); while theta and alpha bands changed coherence only in stages N2 and N3 in three patients.

Two patients with unilateral tinnitus, monaurally stimulated on the left, showed consistent decreased coherence in intraspindle frequency and delta band in the temporal electrodes during the N2 stage and also in the delta band in temporal and frontal electrodes during REM sleep. The only patient with tinnitus on the right showed significant coherence decrease of the spindles in temporal areas during stage N2 (Fig. 7.7, insert). However, some patients with bilateral tinnitus also showed interhemispheric coherence changes with the binaural sound stimulation. In summary, changes in interhemispheric coherence differed depending on the stimulated side and the functional state of the brain, that is, the sleep stage. We cannot rule out any other unknown factors.

CONCLUSIONS

We postulate that the treatment used in this study is effective due to: (1) the use of a highly personalized sound stimulus and (2) the effectiveness of sound stimulation during sleep.

The two clinical trials demonstrated that stimulus treatment with a sound that mimics the characteristics of tinnitus and applied during nighttime sleep resulted in a reduction in the intensity of objectively quantifiable tinnitus in dB (14 dB SPL on average). On the other hand, we showed that the reduction of tinnitus intensity had a direct impact on the improvement of patients' quality of life. The 'physical correlate theory of the sensory scale' states that the mean loudness intensity is equal to half the sound pressure level (SPL −6 dB) (Warren, 1973). Accordingly, the reduction of 14.1 dB achieved implies a significant decrease in perception, since a reduction of 10 dB decreases the perceived intensity of a sound in half (Plack, 2005), the level of perceived tinnitus shows a reduction mean of 62% compared to pretreatment values.

Other authors already postulated that tinnitus was generally developed within frequencies that were deficient in sensory input and that stimulation was more effective if performed in the range of these frequencies (Schaette et al., 2010; König et al., 2006). An additional advantage is the relaxing effect that all patients had to listen to their tinnitus from an external sound source. On the other hand, having the sound created, the patients could show other people how the sound they are constantly listening to and how intense they are, which made them better understood within their environment.

Up to now, all stimulation protocols aimed at neural plasticity in order to adapt the patient to tinnitus, increase tolerance to it, decrease the affective-emotional component triggered by the symptom and control the insomnia produced, in short, reduce the impact on the quality of life with varied tools, in addition to using external sounds to mask tinnitus. Our treatment aims at correcting neural circuits that are aberrantly discharging to the starting point of an alteration in the input of information that the central pathways do not interpret properly. To do this we use the time in which the patient is asleep, based on the antecedents that during sleep the auditory information is processed, that is, the incoming information is reorganized.

Stimulation during sleep is beneficial because of all that has already been argued about auditory processing with the possibility of new learning and generation of memories. Added to this is the advantage that the patients do not have to pay attention to their tinnitus during the day. Furthermore, the treatment does not interfere with daily activities.

About the mechanisms involved in auditory processing during sleep, the main results obtained in the study of power spectrum and coherence of brain waves showed that the major changes occur in stages N2 and N3. The delta and theta bands were the most influenced by the stimulation as well as the spindles during stage N2. In animal models we have described along the entire pathway the temporal interaction between auditory discharges and theta hippocampal rhythm (Velluti and Pedemonte, 2010). All changes were more frequent in temporal areas and more involved with auditory processing than in frontal association areas. Coherence analysis, which is used to evaluate the functional relationship between cortical regions and to quantify the degree of cortical synchronization for certain frequency bands (Cantero et al., 2000), showed the coexistence of frontal and temporal changes, which is not surprising, since it exists as a functional synergy between these areas. Changes in the temporal and frontal regions were also found in deaf patients with intracochlear implants during sleep (Velluti et al., 2010).

Differences between the two hemispheres did not depend, at least exclusively, on the side where the patient perceives tinnitus and hence on the stimulated side, other factors such as brain dominance may be acting in these complex processing processes. These results demonstrate that stimulation with sound during sleep influences brain activity in patients with tinnitus and that this influence is different depending on the stage of sleep being considered.

The results showed a way to objectively reduce the intensity of tinnitus and concomitantly improve the quality of life of the patients. The finding of changes during the sleep of cortical electrophysiology is an approach to investigate the mechanism underlying the reduction of tinnitus intensity and, therefore, to continue improving the treatment protocol.

ACKNOWLEDGEMENT

These works received partial financial support from the Cedars Sinai Medical Center, Los Angeles.

REFERENCES

Andersson, G., Lyttkens, L., Hirvela, C., Furmark, T., Tillfors, M., Fredrikson, M., 2000. Regional cerebral blood flow during tinnitus: a PET case studied with lidocaine and auditory stimulation. Acta Oto-Laringologica 120 (8), 967–972.

Asplund, R., 2003. Sleepiness and sleep in elderly persons with tinnitus. Arch. Gerontol. Geriatr. 37 (2), 139–145.

Ball, W.A., Morrison, A.R., Ross, R.J., 1989. The effects of tones on PGO waves in slow wave sleep and paradoxical sleep. Exp. Neurol. 104, 251–256.

Brozoski, T.J., Bauer, C.A., Caspary, D.M., 2002. Elevated fusiform cell activity in the dorsal cochlear nucleus of chinchillas with psychophysical evidence of tinnitus. J. Neurosci. 22, 2383–2390.

Cantero, J.L., Atienza, M., Salas, R.M., 2000. Clinical value of EEG coherence as electrophysiological index of cortico-cortical connections during sleep. Rev. Neurol. 31 (5), 442–454.

Cipolli, C., 2005. Sleep and memory. In: Parmeggiani, P.L., Velluti, R.A. (Eds.), The Physiologic Nature of Sleep. Imperial College Press, London, pp. 601–629.

Cutrera, R., Pedemonte, M., Vanini, G., Goldstein, N., Savorini, D., Cardinali, D.P., et al., 2000. Auditory deprivation modifies biological rhythms in the golden hamster. Arch. Ital. Biol. 138, 285–293.

De Gennaro, L., Ferrara, M., Bertini, M., 2000. The spontaneous K-complex during stage 2 sleep: is it the "forerunner" of delta waves? Neurosci. Lett. 291, 41–43.

Diekelmann, S., Born, J., 2010. The memory function of sleep. Nat. Rev. Neurosci. 11, 114–126.

Drexler, D., López-Paullier, M., Rodio, S., González, M., Geisinger, D., Pedemonte, M., 2016. Impact of reduction of tinnitus intensity on patients' quality of life. Int. J. Audiol. 55 (1), 11–19.

Drucker-Colín, R., Arankowsky-Sandoval, G., Prospéro-García, O., Jiménez-Anguiano, A., Merchant, H., 1990. The regulation of REM sleep: some considerations on the role of vasoactive intestinal peptide, acetylcholine, and sensory modalities. In: Mancia, M., Marini, G. (Eds.), The Diencephalon and Sleep. Raven Press, New York, pp. 313–330.

Eggermont, J.J., Komiya, H., 2000. Moderate noise trauma in juvenile cats results in profound cortical topographic map changes in adulthood. Hear. Res. 142, 89–101.

Eggermont, J.J., Roberts, L.E., 2004. The neuroscience of tinnitus. Trends. Neurosci. 27, 676–682.

Eysel-Gosepath, K., Selivanova, O., 2005. Characterization of sleep disturbance in patients with tinnitus. Laryngorhinootologie. 84 (5), 323–327.

Folmer, R.L., Griest, S.E., 2000. Tinnitus and insomnia. Am. J. Otolaryngol. 21 (5), 287–293.

Hallam, R.S., 1996. Correlates of sleep disturbance in chronic distressing tinnitus. Scand. Audiol. 25 (4), 263–266.

Hanley, P.J., Davis, P.B., 2008. Treatment of tinnitus with a customized dynamic acoustic neural stimulus: underlying principles and clinical efficacy. Trends Amplif 12 (3), 210–222.

Heijneman, K.M., De Kleine, E., Van Dijk, P., 2012. A randomized double-blind crossover study of phase-shift sound therapy for tinnitus. Otolaryngol Head Neck Surg (USA) 147 (2), 308–315.

Henry, J.A., 2004. Audiologic assessment. In: Snow, J.B. (Ed.), Tinnitus: Theory and Management. Decker, Lewiston, NY, pp. 220–236.

Henry, J.A., Dennis, K., Schechter, M.A., 2005. General review of tinnitus: prevalence, mechanisms, effects, and management. J. Speech Lang. Hear. Res 48, 1204–1235.

Jastreboff, M.M., 2007. Sound therapies for tinnitus management. Prog. Brain. Res. 166, 435–440.

Jastreboff, P.J., 1990. Phantom auditory perception (tinnitus): mechanisms of generation and perception. Neurosci. Res. 8, 221–254.

Jastreboff, P.J., 1995. Tinnitus as a phantom perception: theories and clinical implications. In: Vernon, J.A., Møller, A.R. (Eds.), Mechanisms of Tinnitus. Allyn & Bacon, Boston, MA, pp. 73–93.

Kleinjung, T., Steffens, T., Londero, A., Langguth, B., 2007. Transcranial magnetic stimulation (TSM) for treatment of chronic tinnitus: clinical effects. Prog. Brain. Res. 166, 359–367.

König, O., Schaette, R., Kempter, R., Gross, M., 2006. Course of hearing loss and occurrence of tinnitus. Hear. Res. 221, 59–64.

Kroener-Herwig, B., Biesinger, E., Gerhards, F., Goebel, G., Verena-Greimel, K., Hiller, W., 2000. Retreining therapy for chronic tinnitus. A critical analysis of its status. Scand. Audiol. 29, 67–78.

Kusatz, M., Ostermann, T., Aldridge, D., 2005. Auditory stimulation therapy as an intervention in subacute and chronic tinnitus: a prospective observational study. Int. Tinnutis J. 11 (2), 163–169.

Langguth, B., Salvi, R., Elgoyhen, A.B., 2009. Emerging pharmacotherapy of tinnitus. Expert. Opin. Emerg. Drugs. 14 (4), 687–702.

Lazeyras, F., Boëx, C., Sigrist, A., 2002. Functional MRI of auditory cortex activated by multisite electrical stimulation of the cochlea. Neuroimage 17, 1010–1017.

López González, M.A., López Fernández, R., 2004. Sequential sound therapy in tinnitus. Int. Tinnitus. J. 10 (2), 150–155.

Maquet, P.A.A., Sterpenich, V., Albouy, G., Dang-bu, T., Desseilles, M., Boly, M., et al., 2005. Brain imaging on passing to sleep. In: Parmeggiani, P.L., Velluti, R.A. (Eds.), The Physiologic Nature of Sleep. Imperial College Press, London, pp. 123–138.

Marshall, L., Helgadottir, H., Mölle, M., Born, J., 2006. Boosting slow oscillations during sleep potentiates memory. Nature. 444, 610–613.

McKinney, C., Hazell, J., Graham, R., 1999. An evaluation of the TRT method. In: Hazell, J. (Ed.), Proceedings of the Sixth International Tinnitus Seminar. THC, Cambridge/London, pp. 99–105.

Melcher, J.R., Sigalovsky, I.S., Guinan Jr, J.J., Levine, R.A., 2000. Lateralized tinnitus studied with functional magnetic resonance imaging: abnormal inferior colliculus activation. J Neurophysiol. 83 (2), 1058–1072.

Mölle, M., Born, J., 2011. Slow oscillations orchestrating fast oscillations and memory consolidation. Prog. Brain. Res. 193, 93–110.

Mölle, M., Bergmann, T.O., Marshall, L., Born, J., 2011. Fast and slow spindles during the sleep slow oscillation: disparate coalescence and engagement in memory processing. Sleep. 34, 1411–1421.

Mulders, W.H., Robertson, D., 2009. Hyperactivity in the auditory midbrain after acoustic trauma: dependence on cochlear activity. Neuroscience. 164, 733–746.

Ngo, H.-V.V., Claussen, J.C., Born, J., Mölle, M., 2013. Induction of slow oscillations by rhythmic acoustic stimulation. J. Sleep. Res. 22, 22–31.

Noreña, A.J., Eggermont, J.J., 2003. Changes in spontaneous neural activity immediately after an acoustic trauma: implications for neural correlates of tinnitus. Hear. Res. 183, 137–153.

Noreña, A.J., Eggermont, J.J., 2005. Enriched acoustic environment after noise trauma reduces hearing loss and prevents cortical map reorganization. J. Neurosci. 25, 699–705.

Pantev, C., Okamoto, H., Teismann, H., 2012. Tinnitus: the dark side of the auditory cortex plasticity. Ann. N. Y. Acad. Sci. 1252 (1), 253–258.

Pedemonte, M., Peña, J.L., Torterolo, P., Velluti, R., 1996. Auditory deprivation modifies sleep in the guinea-pig. Neurosci. Lett. 233, 1–4.

Pedemonte, M., Drexler, D., Pol-Fernandes, D., Bernhardt, V., 2007. Tinnitus treatment with sound stimulation during sleep. Cycle. World Federation of Sleep Medicine Sleep Research Societies, Cairns, Australia. Sleep Biol. Rhythms. 5 (1), 54.

Pedemonte, M., Drexler, D., Rodio, S., Geisinger, D., Bianco, A., Pol-Fernandes, D., et al., 2010. Tinnitus treatment with sound stimulation during sleep. Int. Tinnitus. J. 16, 37–43.

Pedemonte, M., Testa, M., Díaz, M., Suárez-Bagnasco, D., 2014. The impact of sound on electroencephalographic waves during sleep in patients suffering from tinnitus. Sleep Sci. 7, 143–151. Available from: https://doi.org/10.1016/j.slsci.2014.09.011. Open acces, online:.

Pilgramm, M., Rychalik, R., 1999. Tinnitus in der Bundesrepublic Deutschland: Eine representative epidemiologische Studie. HNOAktuel 7, 261–265.

Pineda, J.A., Moore, F.R., Viirre, E., 2008. Tinnitus treatment with customized sounds. Int. Tinnitus. J. 14 (1), 17–25.

Plack, Ch., 2005. The Sense of Hearing. Lawrence Erlbaum Associates Publishers, Mahwah, NJ.

Portas, C., Krakow, K., Allen, P., Joseph, O., Armony, J.L., Frith, C.D., 2000. Auditory processing across the sleep–wake cycle: simultaneous EEG and fMRI monitoring in humans. Neuron 28, 991–999.

Rauschecker, J.P., 1999. Auditory cortical plasticity: a comparison with other sensory systems. Trends Neurosci. 22, 74–80.

Reavis, K.M., Rothholtz, V.S., Tang, Q., Carroll, J.A., Djalilian, H., Zeng, F.G., 2012. Temporary suppression of tinnitus by modulated sounds. J. Assoc. Res. Otolaryngol. 13 (4), 561–571.

de Ridder, D., de Mulder, G., Menovsky, T., Sunaert, S., Kovacs, S., 2007. Electrical stimulation of auditory and somatosensory cortices for treatment of tinnitus and pain. Prog. Brain. Res. 166, 377–388.

Riedner, B.A., Hulse, B.K., Murphy, M.J., Ferrarelli, F., Tononi, G., 2011. Temporal dynamics of cortical sources underlying spontaneous and peripherally evoked slow waves. Prog. Brain. Res. 193, 201–218.

Robertson, D., Irvine, D.R.F., 1989. Plasticity of frequency organization in auditory cortex of guinea pigs with partial unilateral deafness. J. Comp. Neurol. 282, 456–471.

Schaette, R., König, O., Hornig, D., Gross, M., Kempter, R., 2010. Acoustic stimulation treatments against tinnitus could be most effective when tinnitus pitch is within the stimulated frequency range. Hear. Res. 269, 95–101.

Tononi, G., Cirelli, C., 2006. Sleep function and synaptic homeostasis. Sleep. Med. Rev. 10, 49–62.

Vallet, M., 1982. La perturbation du sommeil par le bruit. Soz. Praventivmed. 27, 124–131.

Velluti, R.A., 1997. Interactions between sleep and sensory physiology. A review. J. Sleep. Res. 6, 61–77.

Velluti, R.A., 2008. The auditory system in sleep. Elsevier-Academic Press, Amsterdam.

Velluti, R.A., Pedemonte, M., 2010. Auditory neuronal networks in sleep and wakefulness. Int. J. Bifurcat. Chaos 20 (2), 403–407.

Velluti, R.A., Pedemonte, M., 2012. Sensory neurophysiologic functions participating in active sleep processes. Sleep Sci 5 (4), 103–106.

Velluti, R.A., Pedemonte, M., Suárez, H., Bentancor, C., Rodriguez-Servetti, Z., 2010. Auditory input modulates sleep: an intra-cochlear implanted human model. J. Sleep. Res. 19 (4), 585–590.

Vermeire, K., Heyndrickx, K., De Ridder, D., Van De Heyning, P., 2007. Phase-shift tinnitus treatment: an open prospective clinical trial. B-ENT 3 (7), 65–69.

Warren, R.M., 1973. Quantification of loudness. Am. J. Psychol. 86 (4), 807–825.

Wazen, J.J., Daugherty, J., Pinsky, K., Newman, C.W., Sandridge, S., Batista, R., et al., 2011. Evaluation of a customized acoustical stimulus system in the treatment of chronic tinnitus. Otol. Neurotol. 32 (4), 710–716.

von Wedel, H., von Wedel, U.C., Streppel, M., Walger, M., 1997. Effectiveness of partial and complete instrumental masking in chronic tinnitus. Studies with references to retraining therapy. HNO 45 (9), 690–694.

Wilde, R.A., Steed, L., Hanley, P.J., 2008. Treatment of tinnitus with a customized acoustic neural stimulus: a controlled clinical study. Ear Nose Throat J. 87 (6), 330–339.

CHAPTER 8

Other Sensory Modalities in Sleep

Sensory input encompasses a huge spectrum of information reaching the central nervous system (CNS), which generates, after complex processing, expressions of the system in the form of motor acts, endocrine, autonomic, and behavioural responses and changes in the capabilities of the CNS, such as memory and learning. Information from two worlds, the outside and inside (the body), throughout life influences brain development and in particular in the organization of sleep. The development of each brain is genetically determined, although conditioned to very significant events, as is the continued sensory input, a phenomenon that probably begin in uterus and continues throughout life. One consequence of this continuous sensory input to the CNS is that processing will be differential according to the various states of the system, that is, wakefulness and sleep at different levels and different stages (epigenetic factors). An important concept must be added: The CNS can influence the input, since it has efferent control (present in all systems), which can exert influence on the receptors and the various nuclei of the afferent pathway to the sensory cortices. Using this possibility of feedback, sensory processing is complete, because the CNS can select the information that runs a closed circuit (closed loop), selecting entries (Horikawa et al., 2013).

VISION AND SLEEP

Centrifugal fibres that reach the retina were described by Ramón y Cajal (1952) and later shown that the efferent system has direct effects on retinal ganglion cells. However, morel recently Galambos et al. (1994) demonstrated that rat electroretinography was modified during the sleep—wake cycle, with increased amplitude during slow wave sleep (SWS) diminishing similar to waking in the paradoxical (or rapid eye movement, REM) sleep values. These results mimicked those changes obtained simultaneously in evoked potentials in the visual cortex, where they recorded the highest amplitudes during slow sleep.

The Auditory System in Sleep
DOI: https://doi.org/10.1016/B978-0-12-810476-7.00008-7

Continuous exposure to bright light in human subjects introduced changes in the organization of sleep. Unlike with the dim light, bright light induced a significant increase of SWS and decrease of wakefulness, while increased levels of alertness (Kohsaka et al., 1995; Livingstone and Hubel, 1981). Eye movements also may follow REM sleep dream images (Leclair-Visonneau et al., 2010). On the other hand, an intact human visual system is essential to synchronize the circadian system; therefore, blind subjects whose lesions include retino-hipothalamic-fibres show circadian alterations (Leclair-Visonneau et al., 2010).

The analysis of neuronal behaviour at the visual cortex in cats, related to sleep, was studied by Evarts (1963). The spontaneous discharge of neurons in the primary visual cortex of cats was recorded during sleep, demonstrating that more of the neurons discharged at higher frequencies during paradoxical (REM) sleep than during slow sleep.

A different approach to vision and sleep was the analysis of visual magnetic fields (magnetoencephalographic, MEG), presented in the wake state and in each sleep stage, which exhibited variations on passing from wakefulness to N1 and N2 in humans. Technically, the subjects' heads were fixed with biomagnetometers over the cortical occipital area and, stimulated with flashes, induced net changes on passing from wakefulness and sleep (Kakigi et al., 2003) as shown in Fig. 8.1.

Lateral Geniculate Nucleus: A Unitary Approach Carried Out in Guinea Pigs

The configuration of the spontaneous discharge cells in the lateral geniculate nucleus is different in waking and sleep. It described a depression in the response during slow sleep. It was demonstrated that, upon awakening, the neuronal firing changed to either decrease or increase, depending on the stimulus applied (Coenen and Vendrik, 1972).

Other changes described in the configuration of neuronal discharges included in the paradoxical sleep (PS) analysis as well as cross-correlation with hippocampus theta rhythm, changing correlation in the different behavioural stages (Gambini et al., 2002). Pedemonte et al. (2005) demonstrated that, during wakefulness, a change in light flash stimulation pattern (stimuli frequency shift and stimuli 'on' and 'off') caused an increment in the theta band power in 100% of the cases and a phase locking of the spikes in 53% of the recorded neurons. During SWS, no consistent changes in the theta power were observed, notwithstanding that 13% of the neurons exhibited phase locking; that is, novelty may induce changes

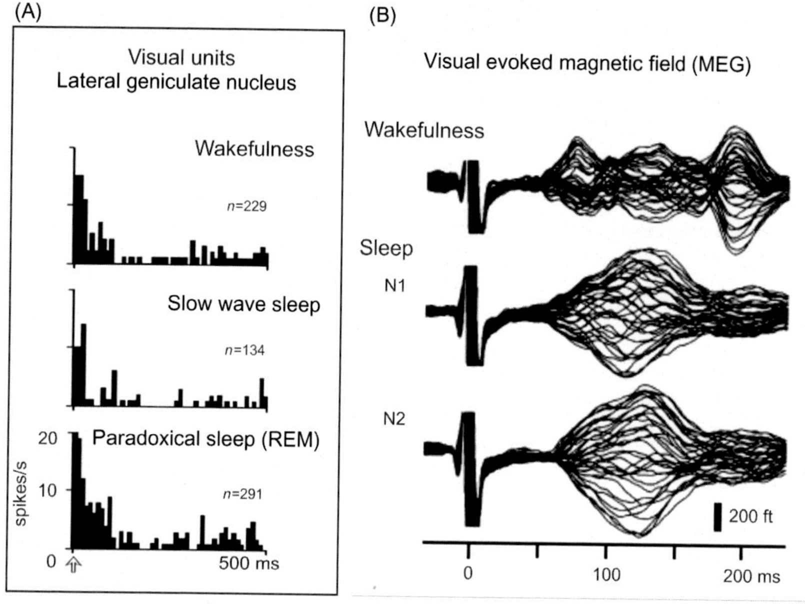

Figure 8.1 Visual bioelectrical variations during wakefulness and sleep. (A) Poststimulus time histograms (*flash*) firing of a visual neuron lateral geniculate nucleus during wakefulness, slow wave sleep, and paradoxical (REM) sleep in a guinea pig. A decrease in the firing number is observed in this case during slow sleep with an increase on passing into PS. (B) Induced magnetic fields (MEG) vision in humans. The waveforms of the responses seen in wakefulness and N1 and N2 sleep with obvious differences. Interindividual differences were observed during wakefulness. Responses during sleep showed similar characteristics, shown in sleep N1 is a peak around 120 ms that increases its amplitude and latency in sleep N2. *Modified from (A) Gambini, J.P., Velluti, R.A., Pedemonte, M., 2002. Hippocampus theta rhythm synchronizes visual neurons in sleep and waking. Brain Res. 926, 37–141 and (B) Kakigi, R., Naka, D., Okusa, T., Wang, X., Inui, K., Qiu, Y., et al., 2003. Sensory perception during sleep in humans: a magnetoencephalograhic study. Sleep Med. 4, 493–507.*

in the temporal correlation of visual neuronal activity with the hippocampus theta rhythm in sleep (Fig. 8.2).

Those results suggest that visual processing in SWS exists, while auditory information and learning were reported during SWS in animals and newborn humans. The changes in the theta power as well as in the neuronal phase locking amount indicate the ability of the hippocampus to contribute to the sensory input analysis in SWS.

Concerns about sleep disturbances in glaucoma patients should be incorporated into clinical evaluations. In addition, glaucomatous patients are potentially associated with other consequences, such as sleep disorders: the

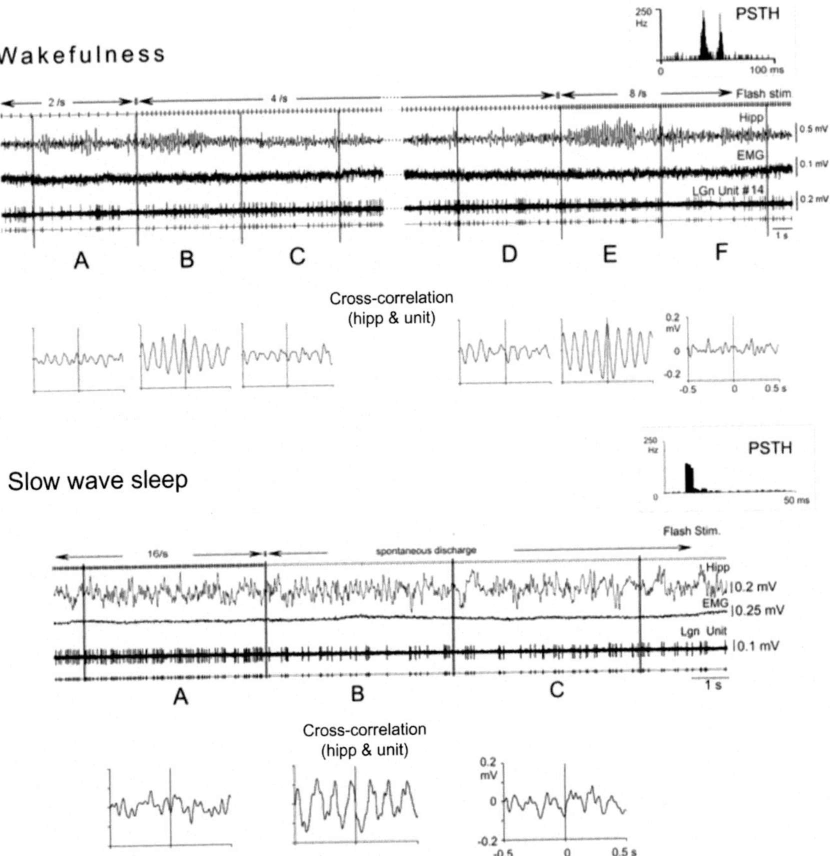

Figure 8.2 Lateral geniculate nucleus (LGn) visual unit discharge and hippocampal theta rhythm pre- and postchanges from frequency stimulation by flashes of light. During wakefulness, at the top, poststimulus time histogram (PSTH) characterizes the unit as a visual one and traces showing from top to bottom: flash stimulation changing from 2/s to 4/s and, after 5 min of recording, shifting from 4/s to 8/s. Hippocampal field activity (hipp), electromyogram (EMG), extracellular unitary discharge (digitized below). This figure shows a quiet wakefulness characteristic recording. Six epochs (5 s each) were selected for processing (A–F, divided by vertical lines). A and D are immediately prior to the frequency stimulation changes, whereas B–C and E–F are successive windows after the light rate stimulation changes. Bottom, each column corresponds to the temporal window shown at the top. Cross-correlation between hippocampal field activity and spikes was calculated by spike-triggered averaging. Temporal correlation (phase locking) between unitary activity and theta rhythm appears after changing the flash rate (windows B and E) and disappears ~5 s later (windows C and F). Hippocampal electrogram power spectra (theta rhythm range shown as black bars) and autocorrelation show an increment in theta rhythm simultaneous with the phase locking.

detect/process novelty, although present, may be decreased. This is consistent with the noticeable decrease in awareness of the environment during sleep (Perl et al., 2016; Gracitelli and Paranhos, 2015).

PROPRIOCEPTIVE SYSTEM

Different objective studies were carried out to evaluate the impact of sleep deprivation on proprioception. The balance perturbations tests, objective posturography measurements and stability were analysed to investigate whether postural stability and adaptation differed after a normal night of sleep compared with results after 24 and 36 hours of sleep deprivation. Sleep deprivation might affect postural stability through reduced adaptation ability and lapses in attention (Patel et al., 2008).

The Vestibular System

Kleitman (1963) observed that nystagmus is greatly reduced or even abolished during sleep. Associated with this, it was demonstrated that the suppression of one receptor by a unilateral labyrinthectomy, in cats, produced spontaneous shift in firing that was present during wakefulness, reduced in SWS, and greatly depressed during PS (REM), during which only irregular eye movements were sometimes interrupted by typical REMs (Baldisera et al., 1967).

◀ During SWS, at the top, PSTH characterizes the unit as a visual one and traces showing from top to bottom are flash synchronizing signal, changing from 16/s to no stimulation (spontaneous activity), hipp, EMG, extracellular unitary discharges (digitized below). This recording is characteristic of SWS because of the high amplitude and low frequency in the hipp field activity and the EMG low activity. Three epochs (5 s each) were selected for processing (A–C, divided by vertical lines); A is immediately previous to the end of flash stimulation while B–C are successive windows without flashes (spontaneous firing). In the lower part, each column corresponds to the temporal window shown above. Cross-correlation between hippocampal field activity and spikes were calculated by spike-triggered averaging. Temporal correlation (phase locking) between unitary discharge and theta rhythm appears after changing the flash stimulation (window B) and disappears some seconds (~5 s) later (window C). Hipp power spectra (theta rhythm range as *black bars*) exhibit two frequency peaks after the flash shift. Hipp wave autocorrelogram shows no dominant frequency after stimulation shift. *Modified from Pedemonte, M., Gambini, J.P., Velluti, R. A., 2005. Novelty-induced correlation between visual neurons and the hippocampus theta rhythm in sleep and wakefulness. Brain Res. 1062, 9–15.*

The trigeminal sensory nuclear complex was recorded with microelectrode analysis of the spontaneous unitary activity during sleep and wakefulness, in unrestrained cats (Badia et al., 1990). Mean rate of second order vestibular unitary discharges was higher during W than during SWS without changes during spindle sleep. In the transition from SWS to PS these cells showed no change or a slight increase in their firing rate. However, when recording form medial and descending vestibular nuclei, a striking effect was observed during a PS phase. The mean rates were two to four times higher during PS than during quiet wakefulness and the pattern of discharge consisted of firing bursts of 80–160 spikes/s, always associated with REM. None of the neurons recorded from vestibular nuclei showed any spontaneous firing shift during eye movements occurring during SWS.

The increases phases in unit discharge must depend on the arrival of extralabyrinthine volleys to the medial and descending vestibular nuclei, since at this time the 'monosynaptic transmission of labyrinthine afferent volleys through the vestibular nuclei is generally depressed during strong arousal reaction as well as during typical REM of paradoxical sleep' (Pompeiano, 1970). These results indicate that a close relation exists between the vestibular nuclei and the bursts of REMs during sleep.

THERMORECEPTION AND SLEEP

Information about body temperature, both increases and decreases, comes from two main sources: the receptors of skin temperature and the central receptors concentrated mainly in the anterior hypothalamus. The hypothalamus integrates information relevant to the regulation of peripheral and core temperature, particularly in view of the mechanisms underlying the powerful influences of temperature changes in the brain over sleep. Exposing animals to variations in ambient temperature, a decrease in SWS and REM sleep were shown. Thermal stress changes, because they promote sleep thermoregulatory responses that are not compatible with the normal development of sleep, for example, long exposure to cold, produce PS deprivation (although without stress control). Neurons in the extrahypothalamic region respond to transient changes in temperature applied to the skin, showing differences in the sleep–wake cycle. Neurons were seen in structures thermosensitive striatal-thalamic-limbic areas located dorsally in relation to classical anterior hypothalamic area; these neurons discharged during wakefulness and SWS, being depressed

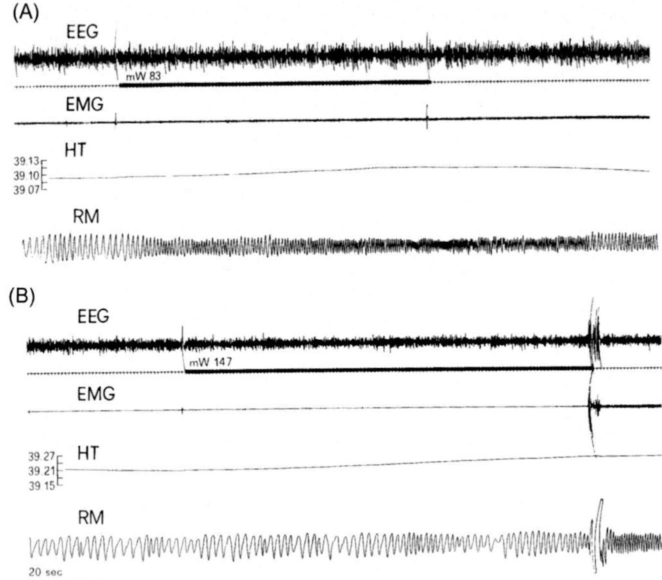

Figure 8.3 Respiratory responses to preoptic-anterior hypothalamic diathermic warming during sleep (cat). (A) Warming elicits tachypnea, which persists after the end of warming in SWS. (B) Warming is ineffective during REMS, notwithstanding the increased intensity. Tachypnea starts immediately with the arousal of the animal. EEG, electroencephalogram; mW, miliwatt; NREMS, NREM sleep, REMS, REM sleep; PG, pneumogram; W, wakefulness. *Modified from Parmeggiani (1973, 2011).*

or altered during REM sleep. It has been shown that, during SWS, cats develop thermoregulatory mechanisms in response to cold and heat; in contrast, these mechanisms are suppressed during PS. The amount of time animals and humans spend in REM sleep depends on the ambient temperature; maximum durations are obtained in neutral temperatures, reducing both the heat and the cold. Variations in ambient temperature cause changes in the sleep architecture (Fig. 8.3, Parmeggiani, 2005, 2011; Libert and Bach, 2005).

The homoeostasis of temperature regulation is different between nap and night sleep, that is, in men sleep does not influence the thermoregulatory system during nap while it does during night sleep (Kräuchi et al., 2006).

OLFACTORY SYSTEM

In the genome of mammals a large percentage of all genes are involved in the detection of odours. The huge amount of genetic information related to smell reflects the significance of this sensory system. For most animals smell is the primary sense used to identify food, predators, colleagues, and so forth, while from the human point of view, smell is mainly an aesthetic sense, without recognizing the many capabilities that this sensory input can have. Little information is available about the olfactory influences on sleep and vice versa. The electrical activity of the olfactory bulb is modulated by the sleep—wake cycle.

Pioneering works demonstrated large waves in the olfactory bulb during wakefulness in cats, characteristic waves in bursts of 40—50 Hz simultaneous with respiration increase during sniffing and decrease during slow sleep, disappearing in the PS, in the cat (Hernández-Peón et al., 1963; Velluti and Hernández-Peón, 1963). A bilateral olfactory bulbectomy in rats significantly reduced the number of episodes of PS, causing an overall reduction in this stage, while not inducing changes in SWS. Furthermore, episodes of PS returned to normal 15 days after the lesion (Araki et al., 1980). Different results were presented: Injury to the lateral portion of the olfactory bulbs in cats caused suppression of sleep. When the olfactory bulbs were electrically stimulated an increase in SWS appeared, while wakefulness decreased. Specific modulation of olfactory input involves the action of centrifugal fibres. The central effects on sensory input are exerted mainly on the mitral cells. Changes in the excitability of mitral cells related to food, effective during wakefulness, disappeared in SWS (Affanni and Cervino, 2005).

Effects of Odours on Sleep

In cats two main groups of fibres from the olfactory bulb have been reported. It is suggested that both are necessary for normal functioning during wakefulness, emphasizing the importance of the medial forebrain bundle in mediating responses during SWS, having shown it takes at least one medial path to have the neocortical awakening in response to the smell of food.

Results in guinea pigs have shown changes in the percentage amounts of sleep stages when income of olfactory information is suppressed by chemical injury of the receptor itself (López et al., 2007; Fig. 8.4).

In humans limited the information about olfaction and sleep is available. The available data indicate the existence of reactions awakening when an olfactory stimulation occurs during sleep (Velluti, 1997). In addition, Perl et al. (2016) found that odour presentation during sleep

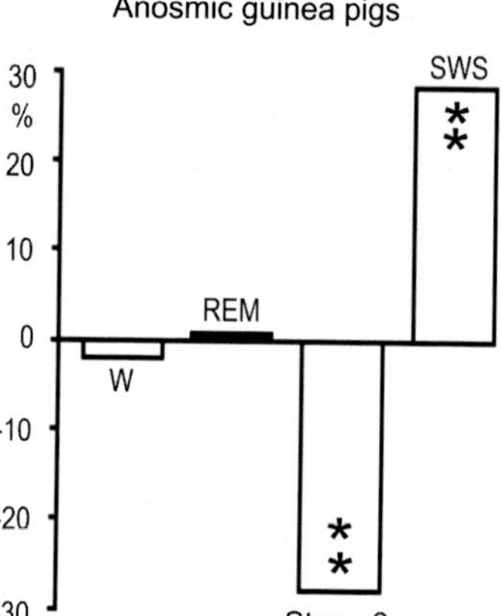

Figure 8.4 Average of percent of changes in sleep stages of anosmic guinea pigs. Olfactory receptors were destroyed through bilateral intranasal instillation of zinc chloride (100–200 mM). Male guinea pigs ($n = 8$) with chronic implants for the diagnosis of sleep–wake cycle were used. Histological monitoring showed total epithelial injury. The results show a statistically significant increment in the percentage of SWS (~30% increase) while sleep spindles decreased (~30%) ($P < 0.001$, Mann–Whitney's test, 95% confidence intervals). Decrease in the percentage of total wake time and increment in the REM were not consistent. *Modified from López, C., Rodríguez-Servetti, Z., Velluti, R.A., Pedemonte, M., 2007. Influence of the olfactory system on the wakefulness-sleep cycle. WFSMSRES World Federation of Sleep Medicine Sleep Research Societies, Cairns, Australia. Sleep Biol. Rhyth. 5, 54.*

enhanced the power of delta (0.5–4 Hz) and slow spindle (9–12 Hz) frequencies during nonrapid eye movement (NREM) sleep. In fact, there is some evidence that several odours promote sleep: In humans lavender oil presented during sleep improved sleep efficiency and increased total sleep time. A woody resinous odour increased the duration of PS in rats (Yamaoka et al., 2005).

SUMMARY

Sensory input and subsequent processing are definitely present in sleep, but show different characteristics than during wakefulness. The interaction between sleep and sensory physiology is an important factor because any sufficiently intense sensory stimulation always produces an awakening, from any stage of sleep.

Interestingly enough, each sensory system has an efferent pathway, with centrifugal projections ending in virtually all core afferents and on the receiver itself. Therefore, incoming sensory information can alter the physiology of sleep and wakefulness, and these states modulate incoming information.

Normal sleep depends on many aspects of sensory input. Neural networks that command sleep and wakefulness are modulated by many sensory inputs, a proportion of the 'passive' effects must be associated with active mechanisms of sleep. Gains or losses sensory inputs produce imbalances in neuronal networks involved in the sleep–wake cycle, changing their relative proportions of active and not being mere passive processes. For example, the almost complete deafferentation in cats caused a state of drowsiness in these animals (Vital-Durand and Michel, 1971).

REFERENCES

Affanni, M., Cervino, J., 2005. Interactions between sleep, wakefulness and the olfactory system. In: Parmeggiani, P.L., Velluti, R.A. (Eds.), The Physiologic Nature of Sleep. Imperial College Press, London, pp. 571–599.

Araki, H., Yamamoto, T., Watanabe, S., Ueki, S., 1980. Changes in sleep wakefulness pattern following bilateral olfactory bulbectomy in rats. Physiol. Behav. 24, 73–78.

Badia, P., Wesenstein, N., Lammers, W., Culpepper, J., Harsh, J., 1990. Responsiveness to olfactory stimuli presented in sleep. Physiol. Behav. 48, 87–90.

Baldisera, F., Broggi, G., Mancia, M., 1967. Nystagmus induced by unilateral labyrinthectomy affected by sleep-wakefulness cycle. Nature 215, 62–63.

Coenen, A.U., Vendrik, A.J.H., 1972. Determination of the transfer ratio of cat's geniculate neurons through quasi-intracellular recordings and the relation with the level of alertness. Exp. Brain Res. 14, 227–242.

Evarts, E.V., 1963. Photically evoked responses in visual cortex units during sleep and waking. J. Neurophysiol. 26, 229–248.

Galambos, R., Juhász, G., Kékesi, A.K., Nyitrai, G., Szilágy, N., 1994. Natural sleep modifies the rat electroretinogram. Proc. Natl Acad. Sci. USA 91, 5153–5157.

Gambini, J.P., Velluti, R.A., Pedemonte, M., 2002. Hippocampus theta rhythm synchronizes visual neurons in sleep and waking. Brain Res. 926, 37–141.

Gracitelli, C.P.B., Paranhos, A., 2015. Can glaucoma affect sleep quality? Arq Bras. Oftalmol. 78 (3), 82–84.

Hernández-Peón, R., Chávez-Ibarra, G., Morgane, J.P., Timo-Iaria, C., 1963. Limbic cholinergic pathways involved in sleep and emotional behavior. Exp. Neurol. 8, 93–111.

Horikawa, T., Tamaki, M., Miyawaki, Y., Kamitani, Y., 2013. Neural decoding of visual imagery during sleep. Science 340, 639–642.

Kakigi, R., Naka, D., Okusa, T., Wang, X., Inui, K., Qiu, Y., et al., 2003. Sensory perception during sleep in humans: a magnetoencephalograhic study. Sleep Med. 4, 493–507.

Kleitman, N., 1963. Sleep and Wakefulness. University of Chicago Press, Chicago, IL, London.

Kohsaka, M., Honma, H., Fukuda, N., et al., 1995. Effect of daytime bright light on sleep structure and alertness. Sleep Res. 24A, 33.

Kräuchi, K., Knoblauch, V., Wirz-Justice, A., Cajochen, C., 2006. Challenging the sleep homeostat does not influence the thermoregulatory system in men: evidence from a nap vs. sleep-deprivation study. Am. J. Physiol. Regul. Integr. Comp. Physiol. 290, 1052–1061.

Leclair-Visonneau, L., Oudiette, D., Gaymard, B., Leu-Semenescu, S., Arnulf, I., 2010. Do the eyes scan dream images during rapid eye movement sleep? Evidence from the rapid eye movement sleep behaviour disorder model. Brain 133, 1737–1746.

Libert, J.P., Bach, V., 2005. Thermoregulation and sleep in the human. In: Parmeggiani, P.L., Velluti, R.A. (Eds.), The Physiologic Nature of Sleep. Imperial College Press, London, pp. 407–429.

Livingstone, M.S., Hubel, D.H., 1981. Effects of sleep and arousal on the processing of visual information in the cat. Nature 291, 554–561.

López, C., Rodríguez-Servetti, Z., Velluti, R.A., Pedemonte, M., 2007. Influence of the olfactory system on the wakefulness-sleep cycle. WFSMSRES World Federation of Sleep Medicine Sleep Research Societies, Cairns, Australia. Sleep Biol. Rhyth. 5, 54.

Parmeggiani, P.L., 1973. The physiological role of sleep. In: Leven, P., Koella, W.P. (Eds.), Sleep. Karger, Basel, pp. 210–216.

Parmeggiani, P.L., 2005. Sleep behavior and temperature. In: Parmeggiani, P.L., Velluti, R.A. (Eds.), The Physiologic Nature of Sleep. Imperial College Press, London, pp. 387–405.

Parmeggiani, P.L., 2011. Systemic Homeostasis and Poikilostasis in Sleep. Imperial Collegge Press, London, pp. 39–78.

Patel, M., Gomez, S., Berg, S., Almbladh, P., Lindblad, J., Petersen, H., et al., 2008. Effects of 24-h and 36-h sleep deprivation on human postural control and adaptation. J. Exp. Brain Res. 85 (2), 165–173.

Pedemonte, M., Gambini, J.P., Velluti, R.A., 2005. Novelty-induced correlation between visual neurons and the hippocampus theta rhythm in sleep and wakefulness. Brain Res. 1062, 9–15.

Perl, O., Arzi, A., Sela, L., Secundo, L., Holtzman, Y., Samnon, P., et al., 2016. Odors enhance slow-wave activity in non rapid eye movement sleep. J. Neurophysiol. 115, 2294–2302.

Pompeiano, O., 1970. Mechanisms of sensorimotor integration during sleep. In: Stellar, E., Sprague, JM. (Eds.), Progress in Physiological Psychology. Academic Press, New York/London, pp. 1–179.

Ramón y Cajal, S. 1952. Histologie du système Nerveux. Consejo Superior de Investigaciones Científicas. Madrid, Spain.

Velluti, R.A., 1997. Interactions between sleep and sensory physiology. A review paper. J. Sleep Res. 6, 61–77.

Velluti, R.A., Hernández-Peón, R., 1963. Atropine blockade within a cholinergic hypnogenic circuit. Exp. Neurol. 8, 20–29.

Vital-Durand, F., Michel, F., 1971. Effets de la desafferentation periphérique sur le cicle veille-sommeil chez le chat. Arch. Ital. Biol. 109, 166–186.

Yamaoka, S., Teruyo Tomita, T., Yoshie Imaizumi, Y., Watanabe, K., Hatanaka, K., 2005. Effects of plant-derived odors on sleep–wakefulness and circadian rhythmicity in rats. Chem. Senses 30 (suppl_1), 264–i265.

CHAPTER 9

Brain Networks and Sleep Generation: A Hypothesis on Neuronal Cell and Assembly Shifts, a New Short Approach

The concept of neuronal networks is defined by the temporally correlated neuronal firing associated for some functional aim, which is the most likely information code, the ensemble coded by cell assemblies. Neuronal groups connected with several other neurons, or groups, can carry out functional cooperation and integration among widely distributed cells, even with different functional properties to subserve a new state or condition. On the other hand, an individual neuron receives more than thousands of synaptic contacts on its membrane, turning its activity into a continuous membrane potential fluctuation, which determines a very unstable physiological condition to constitute a basic code for information processing. Furthermore, the neuronal network/cell assembly may enhance the selective synaptic activity, referring to a dynamic and transient efficacy that we suggest to be correlated to the behavioral dynamic modulation of the sleep process.

Fig. 8.1 shows that lateral geniculate visual neuron activity changes on passing from W to sleep (seen in the poststimulus time histogram, PSTH, left or the visual evoked magnetoencephalographic, MEG, activity, right side) (Velluti and Pedemonte, 2010).

A neuron firing in a functional associated group may process some information and, sometime later, may become associated with other competing and activated neuronal groups for different functional purposes, that is, after passing from wakefulness into sleep. Moreover, it has recently proposed the existence of brain *hub* distributing information on neuronal networks.

The low threshold hub neurons have very extensive axonal arborizations projected over larger distances and make a greater number and stronger synaptic connections than nonhub neurons (Bonifazi et al., 2009).

The Auditory System in Sleep
DOI: https://doi.org/10.1016/B978-0-12-810476-7.00009-9

During, slow wave sleep an auditory cortical neuron exhibited changes in the electroencephalogram and unit firing rate. Furthermore, the neuron discharge is cross-correlated with hippocampus theta rhythm and, sometime later, left side, changed the correlation with hippocampus theta, demonstrating an association shift to a different neuronal network (Fig. 5.13).

At the inferior colliculus, unitary response sound stimuli indicated an active local network, which further shows cross-correlation with theta networks.

At the cortical level (AC), the presence of a cross-correlation with hippocampus theta rhythm network, thus including different networks associated to sleep, while no association was observed just a few seconds later (Fig. 5.13, left).

Moreover, AC cooling inactivates excitatory cortico-fugal pathways in a less activated intrinsic inhibitory network in the inferior colliculus. Cooling of the auditory cortex networks modifies neuronal activity and the network in the inferior colliculus in rats (Fig. 9.1) (Popelar et al., 2016).

Fig. 10.1 shows a different location on the cortical surface of different frequencies of MEG auditory evoked responses. Passing into sleep stage II exhibited a new cortical place (anatomically different) of the MEG evoked response, necessarily demonstrating the shift to another, active network different from that observed during wakefulness.

If we now go to the superior olive (Fig. 5.5) a PSTH is observed during W that presents a net difference when stimulated monaurally or binaurally, clearly differentiating mono- and binaural unit responses. On

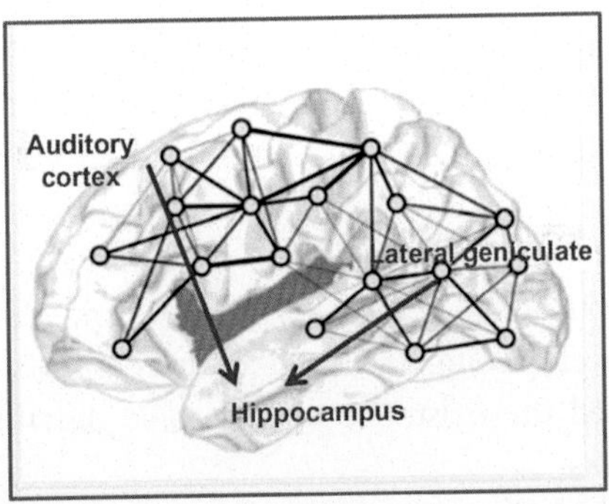

Figure 9.1 An indirect weighted network. The arrows show the relationship between the auditory cortex network generated by sound stimuli and the lateral geniculate nucleus network produced by light stimuli, with the hippocampus theta rhythm network. *Modified from Sporn (2011).*

the other hand, on passing into sleep no difference can be obtained using monaural or binaural stimuli, the PSTH appears totally different, which presents a question: Is the binaural capacity lost in sleep due to the association with another neuronal network?

I believe that there was enough data, auditory as well as visual, to support our hypothesis that sleep is different from waking based on neuronal networks and cell assemblies. A sleep state occurs because there is a shift in all or some cell assemblies or neuronal networks passing from a waking *mode* into a sleeping *mode*, perhaps organized by some unknown *hub(hippocampus)* neuron (Bonifazi et al., 2009).

The interaction between sleep and sensory physiology is an important factor because any sufficiently intense sensory stimulation always produces an awakening, from any stage of sleep. Interestingly enough, each sensory system has an efferent pathway, with centrifugal projections ending in virtually all core afferents and on the receiver itself. Therefore, incoming sensory information can alter the physiology of sleep and wakefulness, and these states modulate incoming information. Normal sleep is in many ways dependent on every sensory input. Neural networks that command sleep and wakefulness are modulated by the sensory inputs, a proportion of the sensory *passive theory* effects on sleep must be associated with sensory *active actions* for sleep organization. Gains or losses in sensory inputs produce imbalances in the neuronal networks involved in the sleep—wake cycle, changing their relative proportions of active stimulation and not being mere passive processes (Velluti and Pedemonte, 2012).

Finally, neuronal networks are easily commanded by the low threshold and highly distributed axonal branches: the *hub* neurons, on which depends the sleep—wakefulness differences.

REFERENCES

Bonifazi, P., Goldin, M., Picardo, M.A., Jorquera, I., Bianconi, G., Represa, A., et al., 2009. GABAergic hub neurons orchestrate synchrony in developing hippocampal networks. Science 326 (5958), 1419—1424. Available from: http://dx.doi.org/10.1126/science.1175509.

Popelar, J., Suta, D., Lindovský, J., Bures, Z., Pysanenko, K., Chumak, T., et al., 2016. Cooling of the auditory cortex modifies neuronal activity inthe inferior colliculus in rats. Hear. Res. 332, 7e16.

Sporn O., 2011. MIT Press. ISBN: 978-0-262-01469.

Velluti, R.A., Pedemonte, M., 2010. Auditory neuronal networks in sleep and wakefulness. Int. J. Bifurcat. Chaos 20, 403—407.

Velluti, R.A., Pedemonte, M., 2012. Sensory neurophysiologic functions participating in active sleep processes. Sleep Sci. 5, 103—106.

CHAPTER 10
Final Conclusions

> *Arriving to the end, some memories about the beginnings an idea appears in my mind, about the first thought devoted to this book subject. Long ago, at the commencement of my research career when looking at an oscilloscope some new auditory collected data, I asked myself where the observed changes were originated in the brain. Besides, what may happen to the data if the brain actually shifts state? Thus, after many experimental approaches and a lot of readings, from the auditory and sleep sides, the plan to analyse together both approaches became real. The first time I have engaged in the study of auditory signals during sleep was in the sixties although it came out as a first regular publication in 1989. R.A. Velluti. CLAE U. Punta del Este Uruguay*
>
> — ***R.A. Velluti. CLAE U. Punta del Este Uruguay***

About 95% of the experimental data reported in this book was carried out in the Laboratorio de Neurofisiología, Facultad de Medicina, Universidad de la República, Montevideo, Uruguay.

The three experimental approaches shown (the auditory neuron's firing, the discharge pattern and the temporal correlation with the hippocampus theta rhythm) represent evidence of sensory processing aspects that occur in sleep and waking. This also gives insight into how sensory information processing and sleep physiology reciprocally affect each other, participating in the processing or in the postulated active promotion of sleep functions (Velluti and Pedemonte, 2010, 2012).

The changes in neuronal discharge rate and pattern in response to constant stimuli indicate that the CNS modulates the incoming auditory information according to the behavioural state, from the auditory nerve to the auditory cortex. Likewise, somatosensory (Pompeiano, 1970; Soja et al., 1998) and visual neurons (Livingstone and Hubel, 1981; Gambini et al., 2002; McCarley, 2004) exhibit changes in their firing rates in correlation with stages of sleep and wakefulness (review Velluti, 1997). This is consistent with the hypothesis that a general shift in the neuronal network/cell assembly's organization is involved in sensory processing that occurs during sleep. This assumption is supported by magnetoencephalographic (MEG) study of auditory stimulation during sleep performed in

The Auditory System in Sleep
DOI: https://doi.org/10.1016/B978-0-12-810476-7.00017-8

humans. It was observed that the dipole location changed in the auditory cortex on passing from wakefulness into slow wave sleep (Fig. 10.1), thus demonstrating the existence of a functional/anatomical network/cell assembly shift in the *planum temporale* area upon passing into sleep (Kakigi et al., 2003; Naka et al., 1999).

A number of neurons at different auditory loci, from the brain stem to the cortex itself, exhibited significant quantitative and qualitative changes in their evoked firing rate and pattern of discharge on passing into sleep. Most important, no neuron belonging to any auditory pathway level or cortex was observed to stop firing during sleep. In addition, our results indicate that the responsiveness of the auditory system during sleep can be considered preserved. Those neurons that continue to fire during sleep equal to their firing during wakefulness (~50% at the primary cortex) are probably related to a continuous monitoring of the environment, whereas the units that increase or decrease their evoked discharge would participate in sleep-related functions, probably associated with different sleep-related active neuronal networks. I cannot advance what their involvement in sleep neurophysiology could be, but it is my hypothesis

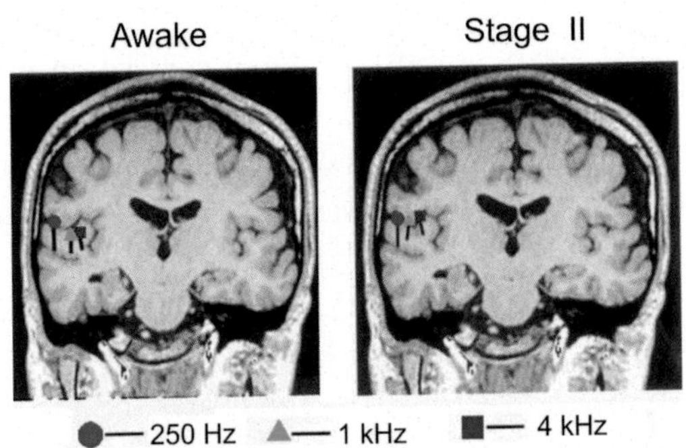

Figure 10.1 *Planun temporale* auditory cortical location of the M100 MEG component observed in response to three sound frequency stimuli (250 Hz, 1 and 4 kHz) recorded in Wakefulness and stage II sleep in humans. The magnetometer was placed on the left hemisphere (C3 position) and the signal source was estimated using an equivalent current dipole (ECD) model. ECD overlapped on MRI. The ECDs (dipoles) were localized deeper in response to the higher frequency tones to the lower frequency tones, while the three changed position on the cortex. The relatively great shifts in the cortical space exhibited by the dipoles demonstrate that the working network changed, surely including new cells elements and communications. *Modified from Naka, D., Kakigi, R., Hoshiyama, M., Yamasaki, H., Okusa, T., Koyama, S., 1999. Structure of the auditory evoked magnetic fields during sleep. Neuroscience 93, 573–583.*

that they are actively involved in sleep processes, such as sleep organization and maintenance, dream content and mechanisms of arousal.

Functional magnetic resonance imaging (MRI) and the evoked potential cortical late waves in humans have also provided evidence that the sleeping brain can process auditory stimuli and detect meaningful events (see Chapter 2: The Physiological Bases of Sleep).

The temporal correlation between hippocampus theta rhythm and the firing of sensory units was shown at several levels, in the auditory pathway (Velluti and Pedemonte, 2002) and in visual neurons of the thalamus (Gambini et al., 2002). At a neuronal population scale, this phase locking may result in a composite final signal that could be used in processes like attention, movement, and auditory sensory incoming information temporal processing. Furthermore, I am proposing that the phase locking to the hippocampus theta waves adds a temporal dimension to the sensory processing, perhaps necessary for time related perception. Every auditory stimulus, every sound, develops in time, that is why the CNS must have a functional possibility to encode this parameter. The hippocampus theta rhythm, being one of the most regular brain-generated low frequency rhythms, may participate as an internal low frequency clock acting as a time giver (Pedemonte et al., 1996, 2001; Velluti et al., 2000; Velluti and Pedemonte, 2002; Pedemonte and Velluti, 2005).

Furthermore, discrimination of significant auditory signals from a background noise is a result of the enhancing excitatory and inhibitory periods in the unit responses to the acoustic stimulus under hippocampus theta influences (Parmeggiani et al., 1982). The phase precession of a CA1 cell and theta waves that occur when a rat is approaching a specific place (that is, the unit appears earlier in relation to theta waves) constitutes an example of temporal coding in the mammalian (Magee, 2003). This precession phenomenon has been also associated with no-spatial behaviours such as during paradoxical sleep (Buszaki, 2002).

The temporal relationship between the neuronal firing and the hippocampus field activity is a varying phenomenon in the time domain that may be dependent on the interaction of a set of signals: the hippocampus rhythm amplitude or frequency; the current state of the brain, awake or asleep; and the characteristics of the incoming sensory information.

The response of auditory neurons to the animal's own vocalizations supports the experimental results obtained by using an artificial sound (tone bursts). In general, the population of cortical neurons stimulated with natural sounds showed wakefulness and slow wave sleep firing shifts and hippocampus theta phase locking in response to artificial stimuli. Furthermore, during slow wave sleep and using natural call stimulus, both responses were present, although with some differences, perhaps representing a component of another processing category in a different neuronal network or cell assembly. In addition, it has been suggested the AI cortical region might serve a general purpose hub of the auditory pathway for the representation of complex sound features to be later complemented with higher auditory centres that further process high level properties (Griffiths et al., 2004).

It is further suggested that the activity-dependent brain development during early life may not only occur during wakefulness (Marks et al., 1995) but also during the long physiological sleep periods of newborns and infants. During early ontogenetic development, and maybe also in adulthood, the sensory information that reaches the CNS during sleep may 'sculpt' the brain and participate in the adaptation to novel conditions.

The initial step towards an auditory learning process is the demonstration that the incoming corresponding information may be differentially processed in sleep. When the auditory information is processed, the person could be ready for learning, perhaps even in sleep. This is consistent with reports of learning during sleep in human newborns (Cheour et al., 2002), consolidation of perceptual learning of spoken language in sleep (Fenn et al., 2003) and visual discrimination improved after sleep (Stickgold et al., 2000) have been observed in different sensory systems (McCarley, 2004).

In addition, the effect of auditory input during sleep of intracochlear implanted patients was studied. Four implanted deaf patients were recorded over four nights: two nights with the implant off, with no auditory input, and two nights with the implant on, that is, with normal auditory input. The sleep patterns of another five deaf people were used as controls, exhibiting normal sleep organization. Moreover, the four experimental patients with intracochlear devices off also showed normal sleep patterns. In comparison with the night recordings with the implant on, a new sleep organization was observed, suggesting that brain plasticity may produce changes in the sleep stage percentages while maintaining the ultradian rhythm. In conclusion, this pilot study shows that the auditory input in humans can introduce changes in central nervous system activity leading to shifts in sleep characteristics, as previously demonstrated in guinea pigs. With the help of a successful intracochlear implant, hearing recovery could produce network reorganization that would consequently alter sleep patterns (Velluti et al., 2010; Velluti and Pedemonte, 2010).

Tinnitus is a sound perception that does not respond to an external sound source. Its prevalence is 10%–15%, affecting the quality of life in 1%–2% of the population. The trigger would be the deregulation of central auditory processing induced by alteration of cochlear entries. Reported results demonstrated that stimulation with sound during sleep influences brain activity in general and particularly in patients with tinnitus. In addition, the sleep influence is different depending on the stage of sleep being considered. The results showed a way to objectively reduce the intensity of tinnitus and concomitantly improve the quality of life of the patients (see Chapter 7: Tinnitus Treatment During Sleep).

REFERENCES

Buszaki, G., 2002. Theta oscillations in the hippocampus. Neuron 33, 325–340.

Cheour, M., Martynova, O., Naatanen, R., Erkkola, R., Sillanpaa, M., Kero, P., et al., 2002. Speech sounds learned by sleeping newborns. Nature 415, 599–600.

Fenn, K.M., Nusbaum, H.C., Margoliash, D., 2003. Consolidation during sleep of perceptual learning of spoken language. Nature 425, 614–616.

Gambini, J.P., Velluti, R.A., Pedemonte, M., 2002. Hippocampal theta rhythm synchronized visual neurones in sleep and waking. Brain Res. 926, 137–141.

Griffiths, T.D., Warren, J.D., Scott, S.K., Nelken, I., King, A.J., 2004. Cortical processing of complex sound: a way forward? Trends Neurosci. 27, 181–185.

Kakigi, R., Naka, D., Okusa, T., Wang, X., Inui, K., Qiu, Y., et al., 2003. Sensory perception during sleep in humans: a magnetoencephalograhic study. Sleep Med. 4, 493–507.

Livingstone, M.S., Hubel, D.H., 1981. Effects of sleep and arousal on the processing of visual information in the cat. Nature 291, 554–561.

Magee, J.C., 2003. A prominent role for intrinsic neuroneal properties in temporal coding. Trends Neurosci. 26, 14–16.

Marks, G.A., Shaffery, J.P., Oksenberg, A., Speciale, S.G., Roffwarg, H.P., 1995. A functional role for REM sleep in brain maturation. Behav. Brain Res. 69, 1–11.

McCarley, R.W., 2004. Mechanisms and models of REM sleep control. Arch. Ital. Biol. 142, 429–467.

Naka, D., Kakigi, R., Hoshiyama, M., Yamasaki, H., Okusa, T., Koyama, S., 1999. Structure of the auditory evoked magnetic fields during sleep. Neuroscience 93, 573–583.

Parmeggiani, P.L., Lenzi, P., Azzaroni, A., D'Alessandro, R., 1982. Hippocampal influence on unit responses elicited in the cat's auditory cortex by acoustic stimulation. Exp. Neurol. 78, 259–274.

Pedemonte, M., Velluti, R.A., 2005. Sleep hippocampal theta rhythm and sensory processing. En: Sleep and Sleep Disorders: a Neuropsychopharmacological Approach. In: Lander, M., Cardinali, D.P., Perumal, P. (Eds.), Landes Biosciencies (Tx). Springer (NY), New York, pp. 8–12.

Pedemonte, M., Peña, J.L., Torterolo, P., Velluti, R.A., 1996. Auditory deprivation modifies sleep in the guinea-pig. Neurosci. Lett. 223, 1–4.

Pedemonte, M., Pérez-Perera, L., Peña, J.L., Velluti, R.A., 2001. Sleep and wakefulness auditory processing: cortical units vs. hippocampal theta rhythm. Sleep Res. Online 4, 51–57.

Pompeiano, O., 1970. The neurophysiological mechanisms of the postural and motor events during desynchronized sleep. Proc. Assoc. Res. Nerve. Ment. Dis. 45, 351–423.

Soja, P.J., Cairns, B.E., Kristensen, M.P., 1998. Transmission through ascending trigeminal and lumbar sensory pathways: dependence on behavioral state. In: Lydic, R., Baghdoyan, H.A. (Eds.), Handbook of Behavioral State Control. CRC Press, Boca Raton, pp. 521–544.

Stickgold, R., Malia, A., Maguire, D., Roddenberry, D., O'Connor, M., 2000. Replaying the game: hypnagogic images in normals and amnesics. Science 290, 350–353.

Velluti, R.A., 1997. Interactions between sleep and sensory physiology. A review paper. J. Sleep Res. 6, 61–77.

Velluti, R.A., Pedemonte, M., 2002. In vivo approach to the cellular mechanisms for sensory processing in sleep and wakefulness. Cell Molec. Neurobiol. 22, 501–516.

Velluti, R.A., Pedemonte, M., 2010. Auditory neuronal networks in sleep and wakefulness. Int. J. Bifurcation Chaos 20, 403–407.

Velluti, R.A., Pedemonte, M., 2012. Sensory neurophysiologic functions participating in active sleep processes. Sleep Sci. 5 (4), 103–106.

Velluti, R.A., Pedemonte, M., Peña, J.L., 2000. Reciprocal actions between sensory signals and sleep. Biol. Signals Recept. 9, 297–308.

Velluti, R.A., Pedemonte, M., Suárez, H., Bentancor, C., Rodríguez-Servetti, Z., 2010. Auditory input modulates sleep: an intra-cochlear-implanted human model. J. Sleep Res. 19, 585–590.

INSERT'S INDEX

CHAPTER 1

Binaural fusion and localization
Masakazu Konishi Caltech Pasadena
California, USA (2008)
Neurotrasmitters, anatomy and function of the nuclei of the lateral lemni cus
Miguel Merchan
Universidad de Salamanca
España

CHAPTER 2

Physiological regulation in sleep
2008 Pier Luigi Parmeggiani
Universita di Bologna
Bologna, Italia
Ponto-geniculo-occipital waves and alerting
2008 Adrian R. Morrison
University of Pennsylvania
Philadelphia, USA

It is still possible that learning mechanisms are ascribed to the dynamic, emergent properties of neural ensembles. We have more neurons than proteins, and perhaps the former can carry out a good job without the need of any structural modifications of their already sophisticated connectivity. Why, then, do most neuroscientists prefer to lean on ***neural plasticity*** *rather than on* ***neural functional states****? The most parsimonious answer is that we have collected a huge amount of information about the structure and connectivity of neural tissue at subcellular and molecular levels, and about the anatomical and biochemical rules and pathways maintaining these structures and circuits. In addition, definite behaviour and sensory-motor properties are easily ascribed to specific neural sites. In contrast, our information about brain functioning during learning situations is too constrained by the limitations imposed by electrical recordings from small numbers of neural elements selected out of billions, or by modern mapping techniques dealing with electrical or biochemical representations of brain activity.*

Jose Maria Delgado, 2008

CHAPTER 3

Cell assemblies and neural networks

Eduardo Mizraji
Facultad de Ciencias
Universidad de la República
Montevideo, Uruguay

CHAPTER 4

No Insert

CHAPTER 5

Brain's maths

José L. Peña, MD, PhD
Dominick P Purpura Department of Neuroscience
Albert Einstein College of Medicine
New York, USA

Rythms and auditory neuronal activity

Marisa Pedemonte, MD, PhD
Universidad CLAEH
Punta del Este, Uruguay (2008)

Brain plasticity versus brain homeostasis

José M. Delgado-García
Universidad Pablo de Olavide
Sevilla, España

CHAPTER 6

*...**Sleep appears to be a behavioural state resulting from dynamic interactions of different physiologic functions** in response to several endogenous (feeding, fatigue, temperature, instinctive drives) and exogenous (light-dark, temperature, food, season, social drives) cues. From the viewpoint of its determination, the mechanism appears complex as to justify a theoretical distinction between **proximate**, **intermediate** and **remote** aspects of determination of sleep behaviour. This gradual approach to sleep behaviour avoids extending the category of rigid causal determination beyond the molecular and cellular levels and forcing experimental results to fit a reductionistic theory in spite of the fact that many elementary physiologic events characterising sleep behaviour are not specific to sleep alone. In other words, sleep, like wakefulness,*

is a function of other interactive functions and not the unique result of the compelling influence of a segregated and highly specific neuronal network of the central nervous system.

Prof. P.L. Parmeggiani, 2008, Universitá di Bologna, Italy

CHAPTER 7

No Insert

CHAPTER 8

No Insert

CHAPTER 9

No Insert

CHAPTER 10

Arriving to the end, some memories about the beginnings come to me, about the first thought devoted to this book subject. Long ago, at the commencement of my research career when looking at an oscilloscope some new auditory collected data, I asked myself where the observed changes were originated in the brain. Besides, what may happen to the data if the brain actually shifts state? Thus, after many experimental approaches and a lot of readings, from the auditory and sleep sides, the plan to analyse together both approaches became real. The first time I have engaged in the study of auditory signals during sleep was in the sixties although it came out as a first regular publication in 1989.

R.A. Velluti. CLAE U. Punta del Este Uruguay

INDEX

Note: Page numbers followed by "*f*", "*t*," and "*b*" refer to figures, tables, and boxes, respectively.